Discovery and Research on Aquatic Microorganisms

Discovery and Research on Aquatic Microorganisms

Editors

Anna Poli
Valeria Prigione

MDPI • Basel • Beijing • Wuhan • Barcelona • Belgrade • Manchester • Tokyo • Cluj • Tianjin

Editors
Anna Poli
University of Torino
Italy

Valeria Prigione
University of Torino
Italy

Editorial Office
MDPI
St. Alban-Anlage 66
4052 Basel, Switzerland

This is a reprint of articles from the Special Issue published online in the open access journal *Applied Sciences* (ISSN 2076-3417) (available at: https://www.mdpi.com/journal/applsci/special_issues/discovery_and_research_on_aquatic_microorganisms).

For citation purposes, cite each article independently as indicated on the article page online and as indicated below:

LastName, A.A.; LastName, B.B.; LastName, C.C. Article Title. *Journal Name* **Year**, *Volume Number*, Page Range.

ISBN 978-3-0365-3797-9 (Hbk)
ISBN 978-3-0365-3798-6 (PDF)

Cover image courtesy of Anna Poli

Contents

About the Editors

Anna Poli and **Valeria Prigione** are currently working at the Mycotheca Universitatis Taurinensis, the Fungal Culture Collection of the Department of Life Sciences and Systems Biology, University of Torino. Their main research interests focus on the fungal biodiversity associated with biotic and abiotic marine substrates, mainly in the Mediterranean Sea, on the description of new marine fungal lineages, and on the investigation of the biotechnological potential of these organisms.

Preface to "Discovery and Research on Aquatic Microorganisms"

Aquatic environments, including freshwater and marine ecosystems, raw and treated sewage, sludge, and sediments, are home to a huge variety of microorganisms that mediate the recycling of dissolved organic carbon and recalcitrant substrata into food webs and the atmosphere. Archaea, bacteria, filamentous fungi, and yeasts play a key role in degradation processes, and many of them are used or have the potential to be harnessed in bioremediation. The importance of aquatic microorganisms is in their physiology and behavior: they can sink or float, some are motile, others adhere to a range of biotic and abiotic substrates (e.g. algae, invertebrates, sediments, driftwood), and they can form biofilms on surfaces, remain planktonic, or produce a broad diversity of bioactive compounds.

By gathering a collection of papers focused on microorganisms in the over-cited environments, this Special Issue will improve the current knowledge of aquatic microbial biodiversity.

Anna Poli and Valeria Prigione
Editors

MDPI

Editorial

Special Issue on Discovery and Research on Aquatic Microorganisms

Anna Poli * and Valeria Prigione *

Mycotheca Universitatis Taurinensis, Department of Life Sciences and Systems Biology, University of Torino, Viale Mattioli 25, 10125 Torino, Italy
* Correspondence: anna.poli@unito.it (A.P.); valeria.prigione@unito.it (V.P.)

Citation: Poli, A.; Prigione, V. Special Issue on Discovery and Research on Aquatic Microorganisms. *Appl. Sci.* **2021**, *11*, 11973. https://doi.org/10.3390/app112411973

Received: 13 December 2021
Accepted: 14 December 2021
Published: 16 December 2021

Publisher's Note: MDPI stays neutral with regard to jurisdictional claims in published maps and institutional affiliations.

Introduction

The Special Issue entitled "Discovery and Research on Aquatic Microorganisms" wished to improve our knowledge on microorganisms living in aquatic environments.

Aquatic environments, including freshwater and marine ecosystems, raw and treated sewage, and sludge and sediments host a huge variety of microorganisms that mediate the recycling of dissolved organic carbon and recalcitrant substrata into the food webs and the atmosphere. Archaea, bacteria, microalgae, filamentous fungi and yeasts play a key role in degradation processes and many of them are used or have the potential to be harnessed in bioremediation. The importance of aquatic microorganisms has to be found in their physiology and behavior: they can sink or float, some are motile, others adhere to a range of biotic and abiotic substrates (e.g., algae, invertebrates, sediments, driftwood, etc.), they can form biofilms on surfaces, remain planktonic, or produce a broad diversity of bioactive compounds.

This Special Issue includes four research articles and one review paper dealing with the applicative potential and the biodiversity of aquatic microorganisms.

Amin et al. [1] faced the problem of bacterial disease outbreakes in aquaculture. By knowing that biofilms formation together with antibiotic resistance protects marine pathogens like *Vibrio campbelli* and *Vibrio parahemolyticus* against antimicrobials, they successfully applied and evaluated the antimicrobial properties and the biofilm inhibition activity of silver ion-236 exchanged zeolite (AgZ).

The importance of filamentous fungi was assessed by Poli et al. [2], Romero-Hernández et al. [3], and by Dobretsov et al. [4]. In the first case, the authors, by applying morphological, molecular and phylogenetic analyses, described the novel *Corollospora mediterranea* species complex (CMSC), isolated from different substrates in the Mediterranean Sea. Species affiliated to the genus *Corollospora* are known to produce bioactive compounds and can be possibly exploited as bioremediators of oil spill contaminated beaches, thus indicating a biotechnological potential of the newly introduced species [2].

Following the Deepwater Horizon oil spill, interest in characterizing the fungal diversity in the Gulf of Mexico (GoM) rapidly increased. Romero-Hernández et al. [3] focused on heavy crude oil (HCO) and extra-heavy crude oil, evaluating the ability of fungal strains isolated from deep-sea sediments of the GoM to degrade them. They demonstrated that species of *Alternaria*, *Penicillium* and *Stemphylium*, can use HCO as its sole carbon source, while a strain of *Alternaria* was the only one to grow in the presence of EHCO, displaying excellent degradative properties.

Dobretsov and collaborators [4], painted plastic panels with a copper-based antifouling paint and exposed them to biofouling for months. The authors then identified six fungal isolates (*Alternaria* sp., *Aspergillus niger*, *A. terreus*, *A. tubingensis*, *Cladosporium halotolerans*, and *C. omanense*) from biofilms developed on the surface of the antifouling paint, demonstrating their resistance to copper. Although further investigations are required, these findings can have implications in the biofouling industry.

Finally, Bolivar-Galiano et al. [5], contributed with a review article focused on the phototroph biodiversity reported in monumental fountains. Since many phototrophic organisms that develop on stone material are responsible for biodeterioration, the authors wished to draw up a guide for interested professionals in the field of conservation, providing simplified dichotomous keys for cyanobacteria, green algae, and diatoms.

The articles published in this Special Issue confirm the importance (both positive and negative), the role, and the possible biotechnological exploitation of different organisms in aquatic ecosystems.

Funding: This research received no external funding.

Conflicts of Interest: The authors declare no conflict of interest.

References

1. Amin, Z.; Waly, N.A.; Arshad, S.E. Biofilm Inhibition and Antimicrobial Properties of Silver-Ion-Exchanged Zeolite A against Vibrio spp Marine Pathogens. *Appl. Sci.* **2021**, *11*, 5496. [CrossRef]
2. Poli, A.; Bovio, E.; Perugini, I.; Varese, G.C.; Prigione, V. Corollospora mediterranea: A Novel Species Complex in the Mediterranean Sea. *Appl. Sci.* **2021**, *11*, 5452. [CrossRef]
3. Romero-Hernandez, L.; Velez, P.; Betanzo-Gutierrez, I.; Camacho-Lopez, M.D.; Vazquez-Duhalt, R.; Riquelme, M. Extra-Heavy Crude Oil Degradation by Alternaria sp. Isolated from Deep-Sea Sediments of the Gulf of Mexico. *Appl. Sci.* **2021**, *11*, 6090. [CrossRef]
4. Dobretsov, S.; Al-Shibli, H.; Maharachchikumbura, S.S.N.; Al-Sadi, A.M. The Presence of Marine Filamentous Fungi on a Copper-Based Antifouling Paint. *Appl. Sci.* **2021**, *11*, 8277. [CrossRef]
5. Bolivar-Galiano, F.; Cuzman, O.A.; Abad-Ruiz, C.; Sanchez-Castillo, P. Facing Phototrophic Microorganisms That Colonize Artistic Fountains and Other Wet Stone Surfaces: Identification Keys. *Appl. Sci.* **2021**, *11*, 8787. [CrossRef]

Article

Biofilm Inhibition and Antimicrobial Properties of Silver-Ion-Exchanged Zeolite A against *Vibrio* spp. Marine Pathogens

Zarina Amin [1], Nur Ariffah Waly [2] and Sazmal Effendi Arshad [2,*]

1 Biotechnology Research Institute, Universiti Malaysia Sabah, Kota Kinabalu 88400, Malaysia; zamin@ums.edu.my
2 Faculty of Science & Natural Resources, Universiti Malaysia Sabah, Kota Kinabalu 88400, Malaysia; arifahwaly@gmail.com
* Correspondence: sazmal@ums.edu.my; Tel.: +60-178180618

Featured Application: This study highlights a potential application of ion exchanged zeolite A against marine microbial pathogens and their biofilms.

Abstract: A challenging problem in the aquaculture industry is bacterial disease outbreaks, which result in the global reduction in fish supply and foodborne outbreaks. Biofilms in marine pathogens protect against antimicrobial treatment and host immune defense. Zeolites are minerals of volcanic origin made from crystalline aluminosilicates, which are useful in agriculture and in environmental management. In this study, silver-ion-exchanged zeolite A of four concentrations; 0.25 M (AgZ1), 0.50 M (AgZ2), 1.00 M (AgZ3) and 1.50 M (AgZ4) were investigated for biofilm inhibition and antimicrobial properties against two predominant marine pathogens, *V. campbelli* and *V. parahemolyticus*, by employing the minimum inhibitory concentration (MIC) and crystal violet biofilm quantification assays as well as scanning electron microscopy. In the first instance, all zeolite samples AgZ1–AgZ4 showed antimicrobial activity for both pathogens. For *V. campbellii*, AgZ4 exhibited the highest MIC at 125.00 µg/mL, while for *V. parahaemolyticus*, the highest MIC was observed for AgZ3 at 62.50 µg/mL. At sublethal concentration, biofilm inhibition of *V. campbelli* and *V. parahemolyticus* by AgZ4 was observed at 60.2 and 77.3% inhibition, respectively. Scanning electron microscopy exhibited profound structural alteration of the biofilm matrix by AgZ4. This is the first known study that highlights the potential application of ion-exchanged zeolite A against marine pathogens and their biofilms.

Keywords: biofilms; zeolite A; *Vibrio* spp.; antimicrobials

Citation: Amin, Z.; Waly, N.A.; Arshad, S.E. Biofilm Inhibition and Antimicrobial Properties of Silver-Ion-Exchanged Zeolite A against *Vibrio* spp. Marine Pathogens. *Appl. Sci.* **2021**, *11*, 5496. https://doi.org/10.3390/app11125496

Academic Editors: Anna Poli and Valeria Prigione

Received: 3 April 2021
Accepted: 21 May 2021
Published: 14 June 2021

1. Introduction

Aquaculture farming has been the fastest growing food producing sector in the last few decades and an important industry in many developing countries. However, the industry currently faces a threatening challenge due to the bacterial disease outbreaks resulting in high mortality rates in the aquaculture population [1,2]. This is in part due to extensive use of antibiotics in fish farms leading to antimicrobial resistance in fish pathogens [3–5]. Vibriosis is an important bacterial disease in wild and farmed marine fishes, which results in severe economic loss of more than USD 1 billion [6]. In addition, bacteria from seafood sources have been associated with foodborne outbreaks [7].

The formation of biofilms by marine pathogens further amplifies problems faced in the aquatic systems and the aquaculture industry. Biofilms are self-assembled communities of bacteria embedded in a self-developed extracellular matrix (ECM) and are adherent to abiotic or abiotic surfaces. The ECM contains exopolysaccharides, proteins and extracellular DNA (eDNA) [8–10]. Biofilm formation is induced by genetic factors and influenced by environmental conditions including pH, temperature and the availability of nutrients [9].

The formation of ECM in biofilms presents a physical barrier that allows the bacteria to protect themselves from external disturbances. This further leads to the improvement of their resilience to the external environmental conditions and enhance their virulence to cause disease. Examples of enhanced resilience include resistance against antimicrobial agents, disinfectants and host defense mechanisms, making them more difficult and expensive to treat [11,12].

The extensive use of antibiotics in marine systems coupled with the enhanced resilience of marine pathogens as biofilms leads to a "double whammy" scenario of antimicrobial resistance in the aquaculture industry, resulting in huge economic losses due to high mortality rates of fish and seafood as important commodities in countries with high economic dependency on seafood farming such as South East Asia and Japan. The prevalent challenge has necessitated efforts on new antibacterial materials that can effectively inhibit growth and resistance of aquatic pathogens while minimizing negative impacts on human and animal health and the environment.

V. parahemolyticus is a prevalent food-poisoning bacterium associated with seafood consumption, typically causing self-limiting gastroenteritis and commonly found in temperate and tropical marine and coastal waters globally [13,14]. Vibriosis is a systemic bacterial infection in farmed and wild marine fishes, which is considered to be a profoundly significant problem due to intensive economic losses in aquaculture industry worldwide [15]. *V. campbelli* is as an emerging marine pathogen recently associated with diseased farm shrimps [16].

Several studies have reported on the development of bacterial biofilms in aquatic environments, particularly challenges faced due to the persistence of these biofilms as important sources of infection and disease in the aquaculture industry. The diseases result in huge economic losses due to high mortality rates of seafood as important commodities [17–19]. Several strategies have been evaluated to combat biofilm development in aquaculture systems, and these include seawater ozonation [20], use of probiotics [21] and sanitizers [22]. In a recent minireview by Artunes et al., the strategy of quorum sensing (QS) inhibition was discussed. QS refers to chemical signaling molecules that control biofilm formation in bacteria [23]. Recently, several studies on antibacterial agents for the aquaculture systems have also been explored. These include non-chemotherapeutic methods such as natural therapeutics from plants and immunostimulants [24–26] and by alternative inorganic materials [27]. Additionally, studies on the application of zeolites as an antimicrobial agent have started to gain traction particularly in the field of aquaculture.

Zeolite is a microporous, crystalline aluminosilicate with a framework that is made up of $[SiO_4]^{4-}$ and $[AlO_4]^{5-}$ tetrahedra. The tetrahedra are joined together by sharing their corner oxygen atom [28]. In this framework, positively charged silicon ion was balanced by oxygen ion. While the negative charged alumina remains unbalanced resulting in a negative charge of the total structure of zeolite, and it is balanced by extra framework cations [29]. These cations can be exchanged with any other positive ions in order to modify their properties to match with the desired application.

Although there has been extensive research on the antibacterial activity of ion-loaded zeolite against common bacteria including *Escherichia coli, Bacillus subtilis, Staphylococcus aureus, Klebsiella pneumoniae, Pseudomonas aeruginosa* and wastewater biofilm bacteria [30–33], to our knowledge, however, there is still relatively a small number of studies of the effectivity of ion-loaded zeolite A against marine pathogens, in particular *Vibrio campbellii* and *Vibrio parahaemolyticus*.

This study aimed to investigate antimicrobial and antibiofilm applications of silver-ion-exchanged zeolite A against marine bacterial pathogens *V. parahaemolyticus* and *V. campbelli*.

2. Materials and Methods

2.1. Liquid Ion Exchange of Zeolite A

The initial synthesis of zeolite A from bentonite clay comprised the activation of precursor bentonite clay by thermochemical treatments in HCl, addition of alkaline activators, ageing and crystallization processes as well as characterization by X-ray diffraction (XRD) and scanning electron microscopy (SEM) analysis (data published elsewhere).

For liquid ion exchange, the integration of silver (Ag) ion into zeolite A was performed by using the liquid ion exchange method as described by Demirci et al. [31], with slight modifications. Briefly, 1 g zeolite A was added into 10 mL of four $AgNO_3$ concentrations: 0.25, 0.50, 1.00 and 1.50 M (denoted AgZ1, AgZ2, AgZ3 and AgZ4, respectively) and mixed at 150 rpm for 3 days. The obtained zeolite A were then vacuum filtered, washed with deionized water and dried at 80 °C overnight.

For the metal ion concentration analysis, a Perkin-Elmer Optima 5300DV (Waltham, MA, USA) axial viewing ICP-OES was used. The analytical wavelengths of elements included were: Na 589.592, Ag 328.068 and Cu 327.393 nm. The software used was WinLab32 for ICP. Briefly, 200 mg of AgZ1, AgZ2, AgZ3 and AgZ4 was added into separate tubes containing 14 mL of acid mixture, aqua regia (40% HNO_3 + 60% HCl) and left overnight at room temperature. Four milliliters of aqua regia solution was then added into the tubes and mixed for 30 min at 80 °C before being filtered and diluted prior the ICP-OES analysis.

2.2. Minimum Inhibition Concentration (MIC) Assay of Ag-Exchanged Zeolite A

The minimal inhibitory concentration (MIC) assay of Ag-exchanged zeolite A against *V. parahaemolyticus* and *V. campbellii* was carried out as described [33]. Briefly, frozen stocks of *V. campbellii* and *V. parahaemolyticus* were grown on nutrient agar (NA) supplemented with 2% of NaCl and incubated for 16 h at 37 °C. Cultures were then inoculated into tryptone soy broth (TSB), and absorbances were adjusted accordingly to an initial starting OD_{600} of 0.05. As seen in Table 1, samples AgZ1, AgZ2, AgZ3 and AgZ4 of respective AgZ concentrations were then added. The assay for each sample was tested with two technical replicates.

Table 1. Silver ion content (mg/g) of zeolite of samples.

Metal Ion/Concentration	0.25 M (1)	0.50 M (2)	1.00 M (3)	1.50 M (4)
Silver (AgZ)	80.949 ± 0.389	160.041 ± 1.328	177.481 ± 1.483	190.511 ± 1.989

For each bacterial species, doubling dilutions up to 1/512 of AgZ1, AgZ2, AgZ3 and AgZ4 were firstly carried out. Five hundred microliters of liquid inoculum (OD_{600} adjusted to 0.05) was then added and incubated in a shaking 37 °C incubator for 16 h. One hundred microliters of the culture was then plated and spread on to nutrient agar (NA) plates supplemented with 2% NaCl and further incubated for 16 h at 37 °C with appropriate controls. The minimum inhibitory concentration (MIC) of AgZ1, AgZ2, AgZ3 and AgZ4 was determined by the lowest concentration of samples that did not exhibit bacterial growth.

2.3. Quantitation of Bacterial Biofilm Growth by Crystal Violet Assay

The biofilm growth of AgZ-treated biofilm were quantitated by the crystal violet assay using the methods described by O'Toole [34] with modifications. Briefly, cultures of *V. campbellii* and *V. parahaemolyticus* were incubated for 16 h at 37 °C in tryptone soy broth (TSB) supplemented with 2% NaCl.

For the biofilm assay, the sublethal concentration obtained from the MIC assay from was selected as the starting assay concentration. Briefly, serial two-fold dilutions of the sample were performed in a 96-well microtiter plate (corning) containing tryptone soy broth supplemented with 2% NaCl. Fifty microliters of the overnight cultures (with OD_{600} adjusted to 0.05) were then inoculated into each well, and it was then incubated at 37 °C for 72 h. Following incubation, wells were washed three times with distilled H_2O, desiccated at 50 °C, stained with 1% crystal violet for 30 min and added with 95% ethanol. The absorbance of solubilized dye was then determined at 570 nm (Shimadzu, Spectramax, Kyoto, Japan).

2.4. Scanning Electron Microscopy (SEM) of *V. campbellii* and *V. parahaemolyticus* Biofilms

The cell morphologies of *V. campbellii* and *V. parahaemolyticus* biofilms grown in TSB media and in TSB with AgZ4 were analyzed by SEM. Briefly, 5 mL overnight liquid cultures of *V. campbellii* and *V. parahaemolyticus* were cell fixated according to protocol by Gomes and Mergulhão [35] with modifications. The biofilm samples were then fixed in 5% glutaraldehyde prepared in 0.1 M PBS pH 7.2 at 4 °C for 12 h. Following the fixation process, the samples were dehydrated by introduction into a series of ethanol solution of varying concentration gradients (35, 50, 75, 95 and 2 × 100%). The dehydrated samples were then immersed in HMDS for 10 min. Upon dehydration, the samples were dried overnight and then sputter-coated with platinum before being analyzed by SEM (Hitachi SEM, S 3400N, Tokyo, Japan).

3. Results

The incorporation of silver ion (Ag) into the zeolite A framework (AgZ) was determined by introducing the zeolite sample to four different concentrations of Ag solution AgZ1, AgZ2, AgZ3 and AgZ4. During this process, the potassium and sodium ions that exist in the zeolite framework were exchanged by metal ions. ICP-OES analysis of AgZ1, AgZ2, AgZ3 and AgZ4 is as shown in Table 1.

As seen in Table 1, the highest silver ion content was observed for AgZ4 at 190.511 ± 1.989 mg/g, followed by the lowest concentration for AgZ1 at 80.949 ± 0.389 mg/g.

3.1. Minimum Inhibitory Concentration (MIC) of *V. campbellii* and *V. parahaemolyticus* in AgZ

As seen in Table 2, both *V. campbellii* and *V. parahaemolyticus* showed susceptibility to all four AgZ1-AgZ4 concentrations. For *V. campbellii*, the highest MIC was observed in AgZ4 at 0.1250 mg/mL. For *V. parahaemolyticus*, the highest MIC was observed in AgZ3 at 0.0625 mg/mL. The lowest MIC for both *V. campbellii* and *V. parahaemolyticus* was observed in AgZ1 at 2.0 and 0.5 mg/mL, respectively.

Table 2. Minimum inhibitory concentration (MIC) of silver-ion-loaded zeolite A against *V. campbellii* and *V. parahaemolyticus*.

	MIC Value of Four Different Silver-Ion-Loaded Zeolite A (mg/mL)			
Bacterial Pathogens	AgZ1	AgZ2	AgZ3	AgZ4
V. campbellii	2.0000	0.5000	0.2500	0.1250
V. parahaemolyticus	0.5000	0.1250	0.0625	0.0625

3.2. Quantitation of Bacterial Biofilm Density by Crystal Violet Assay

The AgZ4 MIC concentration for both pathogens were further selected for biofilm inhibition crystal violet assay. Figure 1 demonstrated biofilm inhibition percentages of ion-exchanged zeolite in comparison with the untreated control. As shown in Figure 1, at 570 nm absorbance, biofilm formation of AgZ4 against *V. campbellii* and *V. parahaemolyticus* exhibited up to 60.2 and 77.3% inhibition, respectively.

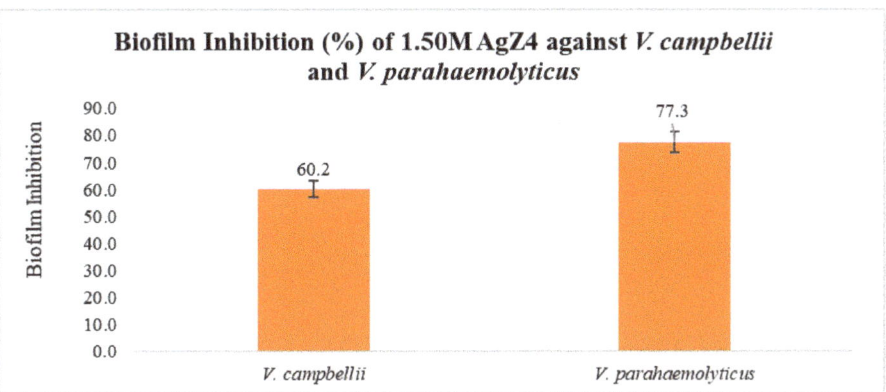

Figure 1. Biofilm inhibition of *V. campbellii* and *V. parahaemolyticus* by 1.50 M AgZ4 assessed by CV staining. Biofilm inhibition of *V. campbellii* and *V. parahaemolyticus* isolates after 24 h growth with 1.50 M AgZ4 was assayed by CV staining (A570).

Scanning Electron Microscopy of *V. campbellii* and *V. parahaemolyticus* in AgZ4, Figures 2a and 3a represent the SEM images of untreated samples of *V. campbellii* and *V. parahaemolyticus*, respectively, while Figures 2b and 3b represent the SEM images of *V. campbellii* and *V. parahaemolyticus* after growth in media culture treated with 1.50 M AgZ4.

Figure 2. SEM analysis showing (**a**) untreated *V. campbellii* biofilm bacteria embedded in extracellular matrix biofilm and (**b**) *V. campbellii* treated with AgZ4 where profound loss of biofilm matrix is seen.

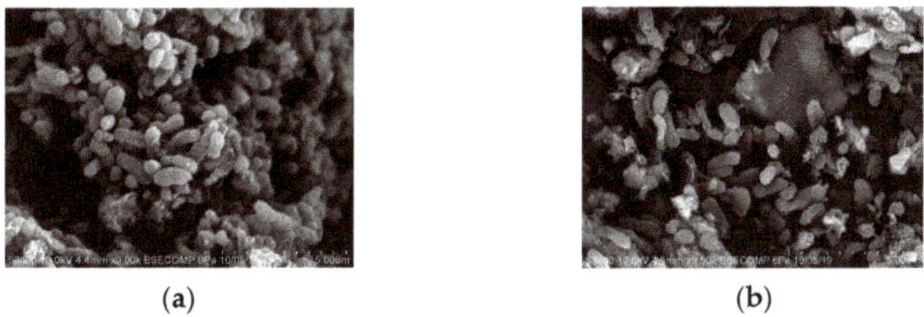

Figure 3. SEM analysis showing (**a**) untreated *V. parahaemolyticus* biofilm bacteria embedded in extracellular matrix biofilm and (**b**) *V. parahaemolyticus* treated with AgZ4 where profound loss of biofilm matrix is seen.

Figure 2a shows clustering and aggregation of *V. campbellii* covered with a network of extracellular matrix (ECM) as a typical representation of intact biofilms. In contrast, Figure 2b shows markedly reduced ECM and aggregation of bacteria.

Similarly, Figure 3a below shows clustering and aggregation of untreated *V. parahaemolyticus* covered with a fine network of extracellular matrix (ECM). In contrast, as seen in Figure 3b, *V. parahaemolyticus* exposure to AgZ4 showed a markedly reduced ECM and cell clustering.

4. Discussion

This study attempted to evaluate the antimicrobial and antibiofilm activities of silver-ion-exchanged zeolite (AgZ) against two marine bacterial pathogens, *V. campbellii and V. parahaemolyticus*. The first stage of this study involved the ICP-OES evaluation of Ag ion incorporation of zeolite synthesized from local bentonite clay (data published elsewhere). The initial ICP-OES analysis of AgZ1, AgZ2, AgZ3 and AgZ4 revealed the successful incorporation of Ag ion into the zeolite samples to be typically in direct correlation with the increase in Ag ion concentration introduced. The highest Ag ion concentration was observed in sample AgZ4 (1.50 M) at 190.511 $\pm$ 1.989 mg/g. At 0.50 M, the Ag ion concentration of AgZ2 was double that of AgZ1 at 160.041 $\pm$ 1.328 mg/g compared to 80.949 $\pm$ 0.389 mg/g. Nonetheless, doubled concentration increases observed for AgZ2 were not indicated in AgZ3 and AgZ4. A possible explanation for this is saturation of available sites in AgZ3 and AgZ4 for the ion exchange to occur.

In general, minimum inhibitory concentration (MIC) evaluation against *V. campbellii* and *V. parahaemolyticus* firstly revealed that the incorporation of Ag ions into the zeolite samples AgZ1, AgZ2, AgZ3 and AgZ4 was successful in inhibiting bacterial growth. Silver (Ag) has long been established in various studies as an effective antibacterial agent. Several studies have also demonstrated the efficiency of Ag-exchanged zeolite against many microbial pathogens [31–33] and concur with the findings of this study. Additionally, a typically inverse relationship between sample concentration and MIC for AgZ1, AgZ2, AgZ3 and AgZ4 was also observed. As seen in Table 2 a lower MIC, which signifies stronger antibacterial activity, was indicated for each bacterial type as the concentration of metal ion in the zeolite increased. Therefore, the higher the Ag ion loading in the zeolite, the lower the concentration of ion-loaded zeolite A needed to inhibit the growth of the bacteria. There have been many studies that have investigated the mechanism of microbial killing by metal ion. A possible mechanism involves the ability of the metal ion to attach to the bacteria membrane through electrostatic interaction and drastically alter the integrity of the bacterial membrane. Consequently, it promotes the formation of reactive oxygen species, (ROS) which will induce the oxidative stress to the bacteria cell resulting in the oxidation of cellular component, DNA damage, mitochondria damage and disruption of the cell membrane, which lead to the death of the bacteria [36–39].

MIC studies also revealed that between the two bacterial species, *V. parahaemolyticus* indicated higher susceptibility against Ag ion compared to *V. campbellii*, as lower MICs were observed across AgZ1–AgZ4. Under antibiotic pressure, bacterial phenotypes such as susceptibility, resistance, tolerance and persistence differ from one bacteria to the other. In a review by Li et al. [40], the efficacy of antimicrobials are influenced by many factors including bacterial status, host factors and antimicrobial concentrations.

A typical feature of bacterial biofilms is the extracellular matrix, which provides protection and structure to the cell population within it. The CV assay is a useful tool for rapid and simple assessment of biofilm formation differences between bacteria, as it stains the extracellular matrix as well as the aggregated bacterial cells [41]. The CV biofilm assays of *V. campbelli* and *V. parahaemolyticus* isolates grown in AgZ4 overnight showed that biofilm growth was effectively inhibited in the presence of AgZ4. However, *V. parahaemolyticus* biofilms were indicated to be more susceptible against AgZ4 with a percentage inhibition of 77.3% compared to *V. campbelli* at 60.2%. This concurred with the MIC assays, which also showed a higher susceptibility of *V. parahaemolyticus* when compared against *V. campbelli* and demonstrated significant attenuation of biofilm formation against *V. campbellii* and *V. parahaemolyticus*.

The ability of Ag ion-exchanged zeolite AgZ4 to disrupt biofilm development for *V. campbellii* and *V. parahaemolyticus* was further supported by SEM analysis. As seen in Figures 2 and 3, the exposure of AgZ4 to both *V. campbellii* and *V. parahaemolyticus* displayed significant structural alteration of biofilm phenotypes when compared to bacterial isolates grown in the absence of AgZ4, including profound loss of the biofilm extracellular matrix (ECM) as well as markedly reduced cell aggregation. While SEM analysis of untreated isolates showed tight aggregation of cells held together by ECM, isolates exposed to AgZ4 showed higher numbers of singular isolates with lessened clustering. Breakages on the extracellular matrices of biofilms will result in increased susceptibility of bacteria against antibacterial agents and chemicals [42]. Despite much literature on antibiofilm activities of bacteria by zeolite, the mechanism of toxicity of AgZ against biofilms of *V. campbellii* and *V. parahaemolyticus* is still poorly understood. Therefore, future studies on the probability of modification of gene expression in the *Vibrio* polysaccharide (VPS) and matrix protein biosynthesis are recommended to further inform on genes or protein that are significantly affected by metal-loaded zeolites.

5. Conclusions

The study findings strongly indicate antimicrobial and antibiofilm characteristics of the silver-ion-exchanged zeolite A against the bacterial pathogens, with the highest MIC levels observed for AgZ4 (1.50 M) for *V. campbelli* and AgZ3 (1.00 M) for *V. parahaemolyticus*. Scanning electron microscopy exhibited profound breakages in the biofilm structures of both marine pathogens when grown in media added with 1.50 M silver-ion-exchanged zeolite A (AgZ4). Taken together, the results of this study strongly indicate the strong antibacterial and antibiofilm potentials of Ag-ion-exchanged zeolite A, which can be applied in the aquaculture industry to combat against infectious pathogens, in particular *V. campbellii* and *V. parahaemolyticus*.

Author Contributions: Formal analysis, Z.A. and N.A.W.; Supervision, S.E.A.; Writing—original draft, Z.A.; Writing—review & editing, Z.A. All authors have read and agreed to the published version of the manuscript.

Funding: This research was funded by Universiti Malaysia Sabah, grant number GUG0111-1/2017. The APC was funded by Universiti Malaysia Sabah.

Informed Consent Statement: Not applicable.

Data Availability Statement: Not applicable.

Conflicts of Interest: The authors declare no conflict of interest.

References

1. Ortega, C.; Mùzquiz, J.; Docando, J.; Planas, E.; Alonso, J. Ecopathology in aquaculture: Risk factors in infections disease outbreak. *Vet. Res.* **1995**, *26*, 57–62.
2. Pridgeon, J.W.; Klesius, P.H. Major bacterial diseases in aquaculture and their vaccine development. *Anim. Sci. Rev.* **2012**, *7*, 1–16. [CrossRef]
3. Miranda, C.D.; Godoy, F.A.; Lee, M.R. Major bacterial diseases in aquaculture and their vaccine development. *Front. Microbiol.* **2018**, *9*, 1–14.
4. Cabello, F.C. Heavy use of prophylactic antibiotics in aquaculture: A growing problem for human and animal health and for the environment. *Environ. Microb.* **2006**, *8*, 1137–1144. [CrossRef]
5. Cabello, F.C.; Godfrey, H.P.; Tomova, A.; Ivanova, L.; Dölz, H.; Millanao, A.; Buschmann, A.H. Antimicrobial use in aquaculture re-examined: Its relevance to antimicrobial resistance and to animal and human health. *Environ. Microb.* **2013**, *15*, 1917–1942. [CrossRef]
6. Zorriehzahra, M.J.; Banaederakshan, R. Early Mortality Syndrome (EMS) as new Emerging Threat in Shrimp Industry. *Adv. Anim. Vet. Sci.* **2015**, *3*, 64–72. [CrossRef]
7. Letchumanan, V.; Yin, W.F.; Lee, L.H.; Chan, K.G.V. Vibrio parahaemolyticus: A review on the pathogenesis, prevalence and advance molecular identification techniques. *Front. Microbiol.* **2015**, *9*, 2513. [CrossRef]
8. Donlan, R.M. Biofilms: Microbial Life on Surfaces. *Emerg. Infect. Dis.* **2002**, *8*, 881–890. [CrossRef]
9. Costerton, J.W.; Stewart, P.S.; Greenberg, E.P. Bacterial biofilms: A common cause of persistent infections. *Science* **1999**, *284*, 1318–1322. [CrossRef]
10. Costa, E.M.; Silva, S.; Tavaria, F.K.; Pintado, M.M. Study of the effects of chitosan upon *Streptococcus* mutans adherence and biofilm formation. *Anaerobe* **2013**, *20*, 27–31. [CrossRef]
11. Maric, S.; Vranes, J. Characteristics and significance of microbial biofilm formation. *Period. Bilogor.* **2007**, *109*, 115–121.

12. Percival, S.L.; Malic, S.; Cruz, H.; Williams, D.W. Introduction to Biofilms. *Biofilms Vet. Med.* **2011**, *14*, 41–68.
13. Mizan, M.F.R.; Jahid, I.K.; Kim, M.; Lee, K.H.; Kim, T.J.; Ha, S.D. Variability in biofilm formation correlates with hydrophobicity and quorum sensing among *Vibrio parahaemolyticus* isolates from food contact surfaces and the distribution of the genes involved in biofilm formation. *Biofouling* **2016**, *32*, 497–509. [CrossRef]
14. Nair, G.B.; Ramamurthy, T.; Bhattachrya, S.K.; Dutta, B.; Takeda, Y.; Sack, D.A. Global dissemination of *V. parahaemolyticus serotype* O3:K6 and its serovariants. *Clin. Microbiol. Rev.* **2007**, *20*, 39–48. [CrossRef]
15. Mohamad, N.; Amal, M.N.A.; Md Yasin, I.S.; Saad, M.Z.; Nasruddin, N.S.; Al saari, N.; Mino, S.; Sawabe, T. Vibriosis in cultured marine fishes: A review. *Aquaculture* **2019**, *512*, 734289. [CrossRef]
16. Haldar, S.; Chatterjee, S.; Sugimoto, N.; Das, S.; Chowdury, N.; Hinenoya, A.; Asakura, M.; Yamasaki, S. Identification of *V. campbelli* isolated from diseased farm shrimps from South India and establishment of its pathogenic potential in an *Artemia* model. *Microbiology* **2011**, *157*, 179–188. [CrossRef] [PubMed]
17. Isiaku, A.I.; Sabri, M.Y.; Ina-Salwany, M.Y.; Hassan, M.D.; Tanko, P.N.; Bello, M.B. Biofilm is associated with chronic streptococcal meningoencephalitis in fish. *Microb. Pathog.* **2017**, *102*, 59–68. [CrossRef]
18. Dang, H.; Lovell, C.R. Microbial Surface Colonization and Biofilm Development in Marine Environments. *Microbiol. Mol. Biol. Rev.* **2016**, *80*, 91–138. [CrossRef]
19. Cai, W.; De La Fuente, L.; Arias, C.R. Biofilm formation by the Fish Pathogen *Flavobacterium columnare*: Development and Parameters Affecting Surface Attachment. *Appl. Environ. Microbiol.* **2013**, *79*, 5633–5642. [CrossRef]
20. Wietz, M.; Hall, M.R.; Hoj, L. Effects of seawater ozonation on biofilm development in aquaculture tanks. *Syst. Appl. Microbiol.* **2009**, *32*, 266–277. [CrossRef]
21. Bentzon-Tilia, M.; Sonnenschein, E.C.; Gram, L. Monitoring and managing microbes in aquaculture. *Microb. Biotechnol.* **2016**, *9*, 576–584. [CrossRef] [PubMed]
22. King, R.K.; Flick, G.J., Jr.; Smith, S.A.; Pierson, M.D.; Boardman, G.D.; Coale, C.W., Jr. Response of Bacterial Biofilms in Recirculating Aquaculture Systems to Various Sanitizers. *J. Appl. Aquac.* **2008**, *20*, 79–92. [CrossRef]
23. Antunes, J.; Leao, P.; Vasconcelos, V. Marine biofilms: Diversity of communities and of chemical cues. *Environ. Microbiol. Rep.* **2018**, *11*, 287–305. [CrossRef]
24. Chakrabarti, R.; Vasudeva, R. *Achyrantes aspera* stimulates the immunity and enhances the antigen clearance in *Catlacatla*. *Int. J. Immunopharmacol.* **2006**, *6*, 782–790. [CrossRef]
25. Albert, V.; Ransangan, J. Antibacterial potential of plant crude extracts against Gram negative fish bacterial pathogens. *Int. J. Res. Pharm. Biosci.* **2013**, *3*, 21–27.
26. Sakai, M. Current research status of fish immunostimulants. *Aquaculture* **1999**, *172*, 63–92. [CrossRef]
27. Thill, A.; Zeynos, O.; Spalla, O.; Chauvat, F.; Rose, J.; Auffan, M.; Flank, A.M. Cytotoxicity of CeO_2 nanoparticles for *Escherichia coli*. Physico-chemical insight of the cytotoxicity mechanism. *Environ. Sci. Technol.* **2006**, *40*, 6151–6156. [CrossRef]
28. Mu, H.; Tang, J.; Liu, Q.; Sun, C.; Wang, T.; Duan, J. Potent antibacterial nanoparticles against biofilm and intracellular bacteria. *Sci. Rep.* **2016**, *6*, 18877. [CrossRef]
29. Breck, D.W. Structure, chemistry and use. In *Zeolite Molecular Sieves*; Wiley: New York, NY, USA, 1974.
30. Belaabed, R.; Elabed, S.; Addaou, A.; Laajab, A.; Rodriguez, M.A.; Lahsini, A. Synthesis of LTA zeolite for bacterial adhesion. *Ceram. Y Vidr.* **2016**, *55*, 152–158. [CrossRef]
31. Demirci, S.; Ustaoğlu, Z.; Yılmazer, G.A.; Sahin, F.; Baç, N. Antimicrobial properties of zeolite-X and zeolite-A ion exchanged with silver, copper, and zinc against a broad range of microorganisms. *Appl. Biochem. Biotech.* **2014**, *172*, 1652–1662. [CrossRef]
32. Zhang, Y.; Zhong, S.; Zhang, M.; Lin, Y. Antibacterial activity of silver-loaded zeolite A prepared by a fastmicrowave loading method. *J. Mater. Sci.* **2009**, *44*, 457–462. [CrossRef]
33. Hajipour, M.J.; Fromm, K.M.; Ashkarran, A.A.; de Aberasturi, D.J.; de Larramendi, I.R.; Rojo, T.; Serpooshan, V.; Parak, W.J.; Mahmoudi, M. Antibacterial properties of nanoparticles. *Trends Biotechnol.* **2012**, *30*, 499–511. [CrossRef] [PubMed]
34. O'Toole, G.A. Microtiter Dish Biofilm Formation Assay. *J. Vis. Exp.* **2011**, *47*, 2437. [CrossRef] [PubMed]
35. Gomes, L.C.; Mergulhão, F.J. SEM analysis of surface impact on biofilm antibiotic treatment. *Scanning* **2017**, *2017*. [CrossRef]
36. Leung, Y.H.; Xu, X.; Ma, A.P.; Liu, F.; Ng, A.M.; Shen, Z.; Gethings, L.A.; Guo, M.Y.; Djurišić, A.B.; Lee, P.K.; et al. Toxicity of ZnO and TiO_2 to Escherichia coli cells. *Sci. Rep.* **2016**, *6*, 35243. [CrossRef] [PubMed]
37. Magana, S.M.; Quintana, P.; Aguilar, D.H.; Toledo, J.A.; Angeles-Chavez, C.; Cortés, M.A.; León, L.; Freile-Pelegrín, Y.; López, T.; Sánchez, R.T. Antibacterial activity of montmorillonites modified with silver. *J. Mol. Catal. A Chem.* **2008**, *281*, 192–199. [CrossRef]
38. Nel, A.E.; Mädler, L.; Velegol, D.; Xia, T.; Hoek, E.M.; Somasundaran, P.; Klaessig, F.; Castranova, V.; Thompson, M. Understanding biophysicochemical interactions at the nano-bio interface. *Nat. Mater.* **2009**, *8*, 543–557. [CrossRef] [PubMed]
39. Krishnani, K.K.; Zhang, Y.; Xiong, L.; Yan, Y.; Boopathy, R.; Mulchandani, A. Bactericidal and ammonia removal activity of silver ion exchanged zeolite. *Bioresour. Technol.* **2012**, *117*, 86–91. [CrossRef]
40. Li, J.; Xie, S.; Ahmed, S.; Wang, F.; Gu, Y.; Zhang, C.; Chai, X.; Wu, Y.; Cai, J.; Cheng, G. Antimicrobial Activity and Resistance: Influencing Factors. *Front. Pharm.* **2017**, *8*, 364. [CrossRef] [PubMed]
41. Haney, E.F.; Trimble, M.J.; Cheng, J.T.; Valle, Q.; Hancock, R.E.W. Critical Assessment of Methods to Quantify Biofilm Growth and Evaluate Antibiofilm Activity of Host Defence Peptides. *Biomolecules* **2018**, *8*, 29. [CrossRef]
42. Dhowlaghar, N.; Bansal, M.; Schilling, M.W.; Nannapaneni, R. Scanning Electron Microscopy of Salmonella biofilms on various food-contact surfaces in catfish mucus. *Food Microbiol.* **2018**, *74*, 143–150. [CrossRef] [PubMed]

Article

Corollospora mediterranea: A Novel Species Complex in the Mediterranean Sea

Anna Poli, Elena Bovio, Iolanda Perugini, Giovanna Cristina Varese * and Valeria Prigione

Mycotheca Universitatis Taurinensis, Department of Life Sciences and Systems Biology, University of Torino, Viale Mattioli 25, 10125 Turin, Italy; anna.poli@unito.it (A.P.); elena.bovio@inrae.fr (E.B.); iolanda.perugini@unito.it (I.P.); valeria.prigione@unito.it (V.P.)
* Correspondence: cristina.varese@unito.com

Abstract: The genus *Corollospora*, typified by the arenicolous fungus *Corollospora maritima*, consists of twenty-five cosmopolitan species that live and reproduce exclusively in marine environments. Species of this genus are known to produce bioactive compounds and can be potentially exploited as bioremediators of oil spill contaminated beaches; hence their biotechnological importance. In this paper, nine fungal strains isolated in the Mediterranean Sea, from the seagrass *Posidonia oceanica* (L.) Delile, from driftwood and seawater contaminated by an oil spill, were investigated. The strains, previously identified as *Corollospora* sp., were examined by deep multi-loci phylogenetic and morphological analyses. Maximum-likelihood and Bayesian phylogeny based on seven genetic markers led to the introduction of a new species complex within the genus *Corollospora*: *Corollospora mediterranea* species complex (CMSC). The Mediterranean Sea, once again, proves an extraordinary reservoir of novel fungal species with a still undiscovered biotechnological potential.

Keywords: marine fungi; new taxa; phylogeny; lignicolous fungi

Citation: Poli, A.; Bovio, E.; Perugini, I.; Varese, G.C.; Prigione, V. *Corollospora mediterranea*: A Novel Species Complex in the Mediterranean Sea. *Appl. Sci.* **2021**, *11*, 5452. https://doi.org/10.3390/app11125452

Academic Editor: Natália Martins

Received: 12 May 2021
Accepted: 11 June 2021
Published: 11 June 2021

Publisher's Note: MDPI stays neutral with regard to jurisdictional claims in published maps and institutional affiliations.

1. Introduction

The last decades have seen an increasing interest in marine fungi due to the need to broaden our knowledge on aquatic biodiversity and to exploit these organisms as a source of novel bioactive molecules. Although more than 1800 species inhabiting the oceans have been described so far, most of the fungal diversity, estimated to exceed 10,000 taxa [1], is yet to be unveiled. Several marine habitats and substrates, both biotic and abiotic, are still being explored worldwide, leading to the discovery of new marine fungal lineages. Among Sordariomycetes, one of the ascomycetous classes mostly detected in the sea, the family Halosphaeriaceae (order Microascales) usually dominates this habitat, with 65 genera and 166 species occurring on driftwood, algae, and seagrasses worldwide [2]. While most of the genera of this family are represented by one or two species, the genus *Corollospora*, typified by *Corollospora maritima* Werderm, includes 25 arenicolous species typically found in beach sand, sea-foam, shell fragments, and algal thalli [3–6]. Besides their ability to rapidly degrade cellulose [7], species affiliated with this genus are known to produce bioactive metabolites [8,9] and can be potentially exploited as bioremediators of oil spill contaminated beaches [10]. For instance, the phthalide derivative corollosporine isolated from *C. maritima* demonstrated a strong antibacterial activity against *Staphylococcus aureus*, *Bacillus subtilis*, and *Escherichia coli* [11], whereas pulchellalactam produced by *C. pulchella*, along with the antimicrobial dioxopiperazines melinacidins II–IV and gancidin W [12], showed inhibitory activity against the tyrosine phosphatase CD45 in lymphocytes [13]. Weak antibacterial activity was also observed in fractions of *C. lacera* mycelial extract [14].

In general, this genus includes morphologically diverse species whose most distinctive features are the ascospore apical primary and equatorial secondary appendages, respectively formed from the epispore and by the fragmentation of the exospore layer [15–18]. Indeed, in reference to the ascospore appendage ontogeny, a revision of the genus was done by scanning and transmission electron microscope investigations [16]. In almost twenty years, other species of *Corollospora* were morphologically described [3,19–24]. With the upcoming molecular techniques, Campbell and collaborators [15] performed phylogenetic analysis of *Corollospora* spp. (and related taxa) based on 28S rDNA sequences and confirmed the monophyletic nature of the genus. However, the authors concluded that more genetic markers, including protein-coding genes, were necessary to resolve relationships among the species of *Corollospora* [15]. New sequence data, including 18S rDNA, ITS, and RPB1, relative to *Corollospora* spp., were then generated in the framework of the Fungal Barcoding Consortium and AFTOL project (Assembly Fungal Tree Of Life) [25,26].

Recently, nine unidentified strains belonging to the genus *Corollospora* were isolated in surveys aimed to investigate the underwater fungal diversity of the Mediterranean Sea: 4 isolates were retrieved from the seagrass *Posidonia oceanica* [27], 4 from seawater contaminated by oil spill [28], and 1 from submerged wood [29]. It is not uncommon, as it is in this case, to come across marine fungi that neither sporulate nor develop reproductive structures in axenic culture, leaving traditional morphology-based identification impossible. Consequently, the identification of these sterile mycelia must rely on molecular data [30–33].

With this study, we wish to provide an accurate phylogenetic placement of the Mediterranean strains by applying a combined multi-locus molecular phylogeny. Besides, the paper gives morphological insights into the strains that turned out to represent new species.

2. Materials and Methods

2.1. Fungal Isolates

The isolates analyzed in this study were previously retrieved from the Mediterranean Sea. In detail, 4 isolates derived from a site chronically contaminated by an oil spill in Sicily (Gela, Italy) [28] 1 from driftwood sampled in the seawater off Porto Badisco (Apulia, Italy) [29] and 4 from leaves of *P. oceanica* collected in good health condition in Tuscany (Elba island, Italy) [27] (Table 1). Originally, the strains were isolated on Corn Meal Agar medium supplemented with sea salts (CMASS; 3.5% *w*/*v* sea salt mix, Sigma-Aldrich, SL, USA, in ddH$_2$O) and are preserved at the Mycotheca Universitatis Taurinensis (MUT), Italy.

Table 1. Dataset used for phylogenetic analysis. Genbank sequences include newly generated nrITS, nrLSU, nrSSU, RPB1, RPB2, TEF-1α, and βTUB amplicons (in bold) relative to the novel species Corollospora mediterranea.

Species	Strain	Source	nrITS	nrSSU	nrLSU	TEF-1α	RPB1	RPB2	βTUB
Microascales									
Halosphaeriaceae									
Corollospora angusta	NBRC 32102	Sea foam	JN943383	JN941667	JN941478	–	JN992397	–	–
	NBRC 32101 T	Seafoam	JN943381	JN941668	JN941477	–	–	–	–
C. cinnamomea	NBRC 32125	Beach sand	AB361023	JN941666	JN941479	–	JN992396	–	–
	NBRC 32126	Beach sand	–	JN941665	JN941480	–	JN992395	–	–
C. colossa	NBRC 32103 T	Seafoam	–	JN941664	JN941481	–	JN992394	–	–
	NBRC 32105	Seafoam	JN943445	JN941663	JN941482	–	–	–	–
C. fusca	NBRC 32107 T	Seafoam	JN943382	JN941662	JN941483	–	JN992393	–	–
	NBRC 32108	Seafoam	JN943385	JN941661	JN941484	–	JN992392	–	–
C. gracilis	NBRC 32110 T	Seafoam	–	JN941660	JN941486	–	JN992390	–	–
	NBRC 32111	Seafoam	JN943386	JN941659	JN941487	–	JN992389	–	–
C. lacera	NBRC 32121	Sea foam	–	JN941658	JN941488	–	JN992388	–	–
	NBRC 32122	Seafoam	–	JN941657	JN941489	–	JN992387	–	–
C. luteola	NBRC 31315 T	Sea foam	–	JN941656	JN941490	–	–	–	–
	NBRC 31316	*Sargassum giganteifolium*	–	JN941655	JN941491	–	–	–	–
C. maritima	MUT 1652	Driftwood	MW543050 *	MW543046 *	MW543054 *	MW577242 *	MW577246 *	MW577250 *	MW556316 *
	MUT 1662	Submerged wood	MW543051 *	MW543047 *	MW543055 *	MW577243 *	MW577247 *	MW577251 *	MW556317 *
	MUT 3408	Submerged wood	MW543052 *	MW543048 *	MW543056 *	MW577244 *	MW577248 *	MW577252 *	MW556318 *
	MUT 3410	Submerged wood	MW543053 *	MW543049 *	MW543057 *	MW577245 *	MW577249 *	MW577253 *	MW556319 *

Table 1. *Cont.*

Species	Strain	Source	nrITS	nrSSU	nrLSU	TEF-1α	RPB1	RPB2	βTUB
Corollospora mediterranea sp. nov.	MUT 1587	Driftwood	KF915998	MW584971 *	MW584962 *	MW703375 *	MW645216 *	MW666025 *	MW727528 *
	MUT 1938	Oil-contaminated seawater	MW582548 *	MW584966 *	MW584957 *	MW703370 *	MW645212 *	n.d.	MW727523 *
	MUT 1950 T	Oil-contaminated seawater	KU935664	MW584972 *	MW584963 *	MW703376 *	MW645217 *	MW666026 *	MW727529 *
	MUT 1954	Oil-contaminated seawater	KU935659	MW584973 *	MW584964 *	MW703377 *	MW645218 *	MW666027 *	MW727530 *
	MUT 1961	Oil-contaminated seawater	KU935662	MW584974 *	MW584965 *	MW703378 *	MW645219 *	MW666028 *	MW727531 *
	MUT 5040	*P. oceanica*	MW582549 *	MW584967 *	MW584958 *	MW703371 *	MW645213 *	MW666022 *	MW727524 *
	MUT 5048	*P. oceanica*	MW582550 *	MW584968 *	MW584959 *	MW703372 *	MW645214 *	MW666023 *	MW727525 *
	MUT 5049	*P. oceanica*	MW582551 *	MW584969 *	MW584960 *	MW703373 *	n.d.	n.d.	MW727526 *
	MUT 5082	*P. oceanica*	MW582552 *	MW584970 *	MW584961 *	MW703374 *	MW645215 *	MW666024 *	MW727527 *
C. pseudopulchella	NBRC 32113	Seafoam	–	JN941651	JN941495	–	JN992383	–	–
	NBRC 32112 T	Seafoam	–	JN941652	JN941494	–	JN992384	–	–
C. pulchella	NBRC 32123	Beach sand	JN943446	JN941650	JN941496	–	JN992382	–	–
	NBRC 32124	Beach sand	–	JN941649	JN941497	–	JN992381	–	–
C. quinqueseptata	NBRC 32114 T	Sea foam	–	JN941648	JN941498	–	JN992380	–	–
	NBRC 32115	*Sargassum sagamianum*	–	JN941647	JN941499	–	JN992379	–	–
Magnisphaera stevemossago	CBS 139776		–	KT278691	KT278704	–	–	KT278740	–
Natantispora unipolaris	NTOU3741		KM624523	KM624521	KM624522	–	–	–	–
Pileomyces formosanus	BBH30192		JX003862	KX686803	KX686804	–	–	–	–
Remispora maritima	BBH28309		–	HQ111002	HQ111012	–	–	–	–
Tinhaudeus formosanus	NTOU3805		KT159895	KT159897	KT159899	–	–	–	–
Microascaceae			–	–	–	–	–	–	–
Cephalotrichum stemonitis	AFTOL-ID 1380	n.d	–	DQ836901	DQ836907	DQ836916	–	–	–
Microascus trigonosporus	AFTOL-ID 914	n.d	DQ491513	DQ471006	DQ470958	DQ471077	DQ471150	DQ470908	–
Petriella setifera	CBS 437.75	Driftwood	–	DQ471020	DQ470969	–	DQ842034	DQ836883	–

* = newly generated sequences; n.d. = not determined; T Type Strain. AFTOL = Assembly Fungal Tree Of Life; BBH = BIOTEC Bangkok Herbarium; CBS = Centraalbureau voor Schimmelculture; MUT = Mycotheca Universitatis Taurinensis; NBRC = Nite Biological Resource Centre; NTOU = National Taiwan Ocean University.

2.2. Morphological Analysis

All isolates were pre-grown on Malt Extract Agar-sea water (MEASW; 20 g malt extract, 20 g glucose, 2 g peptone, 20 g agar in 1 L of seawater) for one month at 21 °C prior to inoculation in triplicate onto new MEASW Petri dishes (9 cm Ø). Petri dishes were incubated at 15 and/or 21 °C. The colony growth was monitored periodically for 28 days, while macroscopic and microscopic features were assessed at the end of the incubation period.

In an attempt to induce sporulation, sterile pieces of *Quercus ruber* cork and *Pinus pinaster* wood (species autochthonous to the Mediterranean area) were placed on 3-week old fungal colonies [34]. Petri dishes were further incubated for 4 weeks at 21 °C. Following, cork and wood specimens were transferred to 50 mL tubes containing 20 mL of sterile seawater. Samples were incubated at 21 °C for at least three months.

Morphological structures were observed, and images captured using an optical microscope (Leica DM4500B, Leica microsystems GmbH, Germany) equipped with a camera (Leica DFC320, Leica microsystems GmbH, Germany).

2.3. DNA Extraction, PCR Amplification, and Data Assembling

Approximately 100 mg of fresh mycelium were carefully scraped from MEASW plates, transferred to a 2 mL Eppendorf tube, and disrupted by the mean of an MM400 tissue lyzer (Retsch GmbH, Haan, Germany). Genomic DNA was extracted following the manufacturer's instructions of a NucleoSpin kit (Macherey Nagel GmbH, Duren, DE, USA). The quality and quantity of DNA were measured spectrophotometrically (Infinite 200 PRO NanoQuant; TECAN, Switzerland); DNA samples were stored at −20 °C.

The partial sequences of seven genetic markers were amplified by PCR. Primer pairs ITS1/ITS4 [35], LR0R/LR7 [36], NS1/NS4 [35] were used to amplify the internal transcribed spacers, including the 5.8S rDNA gene (nrITS), 28S large ribosomal subunit (nrLSU) and 18S small ribosomal subunit (nrSSU). The translation elongation factor (TEF-1α), the β-tubulin (β-TUB) and the largest and second-largest subunits of RNA polymerase II (RPB1 and RPB2) were amplified by using the following primer pairs: EF-dF/EF-2218R [37], Bt2a/Bt2b [38], RPB1Af/RPB1Cr [39] and fRPB2-5F/fPB2-7R [40]. Amplifications were run in a T100 Thermal Cycler (Bio-Rad, Hercules, CA, USA) programmed as described in Table 2. Reaction mixtures consisted of 20–40 ng DNA template, 10× PCR Buffer (15 mM $MgCl_2$, 500 mM KCl, 100 mM Tris-HCl, pH 8.3), 200 μM each dNTP, 1 μM each primer, 2.5 U Taq DNA Polymerase (Qiagen, Chatsworth, CA, USA), in 50 μL final volume. For problematic cases, additional $MgCl_2$, BSA, and/or 2.5% DMSO facilitated the reaction.

Table 2. Genetic markers, primers, and thermocycler conditions were used in this study.

	Forward and Reverse Primers	Thermocycler Conditions	References
ITS	ITS1-ITS4	95 °C: 5 min, (95 °C: 40 s, 55 °C: 50 s, 72 °C: 50 s) × 35 cycles; 72 °C: 8 min; 4 °C: ∞	[35]
LSU	LR0R-LR7	95 °C: 5 min, (95 °C: 1 min, 50 °C: 1 min, 72 °C: 2 min) × 35 cycles; 72 °C: 10 min; 4 °C: ∞	[36]
SSU	NS1-NS4	95 °C: 5 min, (95 °C: 1 min, 50 °C: 1 min, 72 °C: 2 min) × 35 cycles; 72 °C: 10 min; 4 °C: ∞	[35]
TEF-1α	EF-dF/EF-2218R	95 °C: 5 min, (95 °C: 1 min, 50 °C: 1 min; 72 °C: 2 min) × 40 cycles, 72 °C: 10 min; 4 °C: ∞	[37]
βTUB	Bt2a-Bt2b	94 °C: 4 min, (94 °C: 35 sec, 58 °C: 35 s, 72 °C: 50 s) × 35 cycles; 72 °C: 5 min; 4 °C: ∞	[38]
RPB1	RPB1Af-RPB1Cr	96 °C: 5 min, (94 °C: 30 s, 52 °C: 30 s, 72 °C: 1 min) × 40 cycles; 72 °C: 8 min; 4 °C: ∞	[39]
RPB2	fRPB2-5F/fPB2-7cR	94 °C: 3 min, (94 °C: 30 s; 55 °C: 30 s; 72 °C: 1 min) × 40 cycles, 72 °C: 10 min; 4 °C: ∞	[40]

Amplicons, together with a GelPilot 1 kb plus DNA Ladder, were visualized on a 1.5% agarose gel stained with 5 mL 100 mL^{-1} ethidium bromide; PCR products were purified and sequenced at the Macrogen Europe Laboratory (Madrid, Spain). The resulting Applied Biosystem (ABI) chromatograms were inspected, trimmed, and assembled to obtain consensus sequences using Sequencer 5.0 (GeneCodes Corporation, Ann Arbor, MI, USA, http://www.genecodes.com acessed on 12 May 2021). Newly generated sequences were deposited in GenBank (Table 1).

2.4. Sequence Alignment and Phylogenetic Analysis

A dataset consisting of nrSSU, nrITS, nrLSU, and RPB1 was assembled based on BLASTn results and of the most recent phylogenetic studies focused on Halosphaeriaceae and *Corollospora* [30,41]. Reference sequences were retrieved from GenBank (Table 1). Sequences were aligned using MUSCLE (default conditions for gap openings and gap extension penalties), implemented in MEGA X (Molecular Evolutionary Genetics Analysis), visually inspected, and manually trimmed to delimit and discard ambiguously aligned regions. Since no incongruence was observed among single-loci phylogenetic trees, alignments were concatenated into a single data matrix with SequenceMatrix [42]. The best evolutionary model under the Akaike Information Criterion (AIC) was determined with jModelTest 2 [43]. Phylogenetic inference was estimated using Maximum Likelihood (ML) and Bayesian Inference (BI) criteria. The ML analysis was generated using RAxML v. 8.1.2 [44] under GTR + I + G evolutionary model and 1000 bootstrap replicates. Support values from bootstrapping runs (MLB) were mapped on the globally best tree using the "-f a" option of RAxML and "-x 12345" as a random seed to invoke the novel rapid bootstrapping algorithm. BI was performed with MrBayes 3.2.2 [45] with the same substitution model (GTR + I + G). The alignment was run for 10 million generations with two independent runs each containing four Markov Chains Monte Carlo (MCMC) and sampling every 100 iterations. The first 25% of generated trees were discarded as "burn-in". A consensus tree was generated using the "sumt" function of MrBayes and Bayesian posterior probabilities (BPP) were calculated. Consensus trees were visualized in FigTree v. 1.4.2 (http://tree.bio.ed.ac.uk/software/figtree acessed on 12 May 2021). Three species of Microascaceae, namely *Doratomyces stemonitis*, *Microascus trigonosporus*, and *Petriella setifera*, were used to root the tree. Due to the topological similarity of the two resulting trees, only Bayesian analysis with MLB and BPP values was reported (Figure 1).

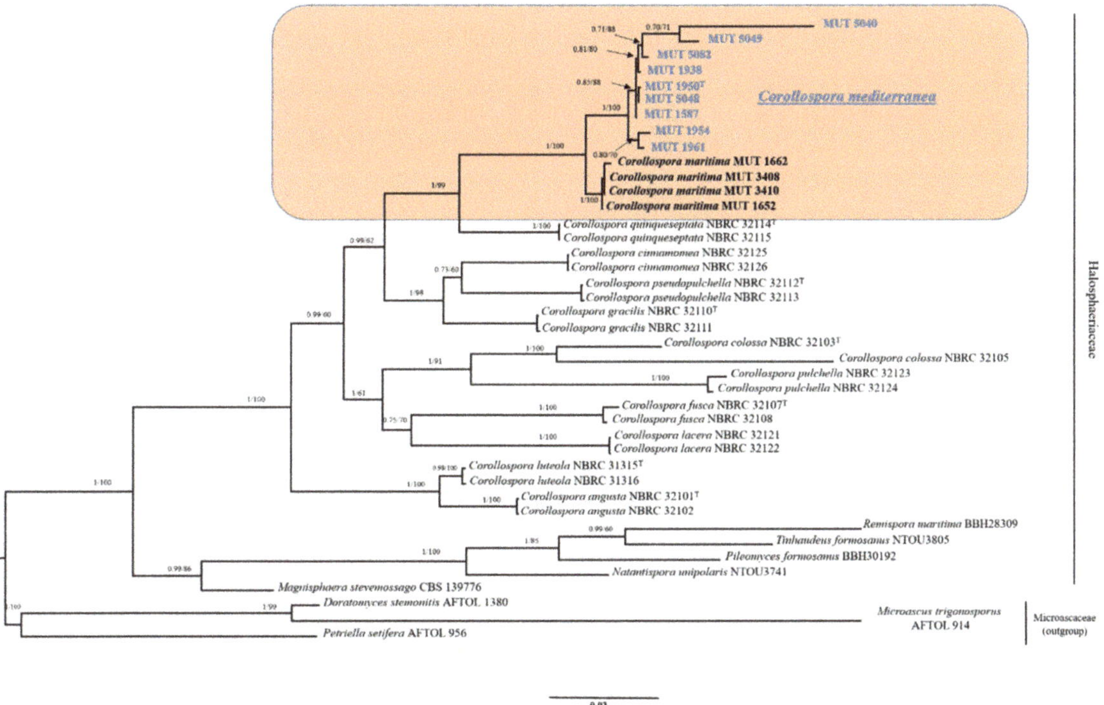

Figure 1. Phylogenetic inference based on a combined nrITS, nrSSU, nrLSU, RPB1 dataset. The tree is rooted in three species of Miroascaceae. Branch numbers indicate BYPP and BS values; T Type Strain; Bar = expected changes per site (0.03).

Following, a new phylogenetic analysis was conducted only on the strains investigated, whose relationship was unclear. To this aim, TEF-1α, β-TUB, and RPB2 sequences were added to the restricted dataset. Alignments and multi-loci phylogeny were conducted as described above. Sequence alignments and phylogenetic trees were deposited in TreeBASE (http://www.treebase.org, submission number S27921 and S27923, acessed on 12 May 2021).

3. Results

3.1. Phylogenetic Inference

Preliminary analyses carried out individually with nrITS, nrSSU, nrLSU, and RPB1 revealed no incongruence in the topology of the single-locus trees. The combined four-markers dataset–built based on BLASTn results and of recent phylogenetic studies [30,41]—Consisted of 42 taxa (including MUT isolates) that represented 9 genera and 20 species (Table 1). A total of 82 sequences (9 nrITS, 13 nrSSU, 13 nrLSU, 13 TEF-1α, 13 β-TUB, 11 RPB1, and 10 RPB2) were newly generated while 91 were retrieved from GenBank.

The dataset combining nrSSU, nrITS, nrLSU and RPB1 had an aligned length of 2431 characters, of which 1530 were conserved, 263 were parsimony-uninformative and 638 parsimony informative (TL = 2192, CI = 0.550901, RI = 0.758483, RC = 0.417849, HI = 0.449099). The strains under investigation, MUT 1587, MUT 1938, MUT 1950, MUT 1954, MUT 1961, MUT 5040, MUT 5048, MUT 5049 and MUT 5082 formed a strongly supported monophyletic lineage (BYPP = 1.00; MLB = 100%) close but well set apart from Corollospora marittima (Figure 1). Within this new group, MUT 1954 and MUT 1961 seemed to form a separated clade, as well as MUT 5040 and MUT 5049. However, the relationships between the taxa were unclear.

The additional dataset, implemented with the addition of TEF-1α, β-TUB, and RPB2 sequence data relative to the novel lineage and to Corollospora marittima only, had an aligned length of 4086 characters, of which 3693 were constant, 236 were parsimony-uninformative and 157 parsimony informative (TL = 452, CI = 0.843537, RI = 0.900217, RC = 0.759367, HI = 0.156463). Two clusters could be observed: the first grouped MUT 1954 and MUT 1961 while the second one was further split into two subclades that separated MUT 1587, MUT 1938, and MUT 5049 from MUT 5040, MUT 5048, and MUT 5082. Finally, MUT 1950, although included second cluster, was not part of any of the two subclades (Figure 2).

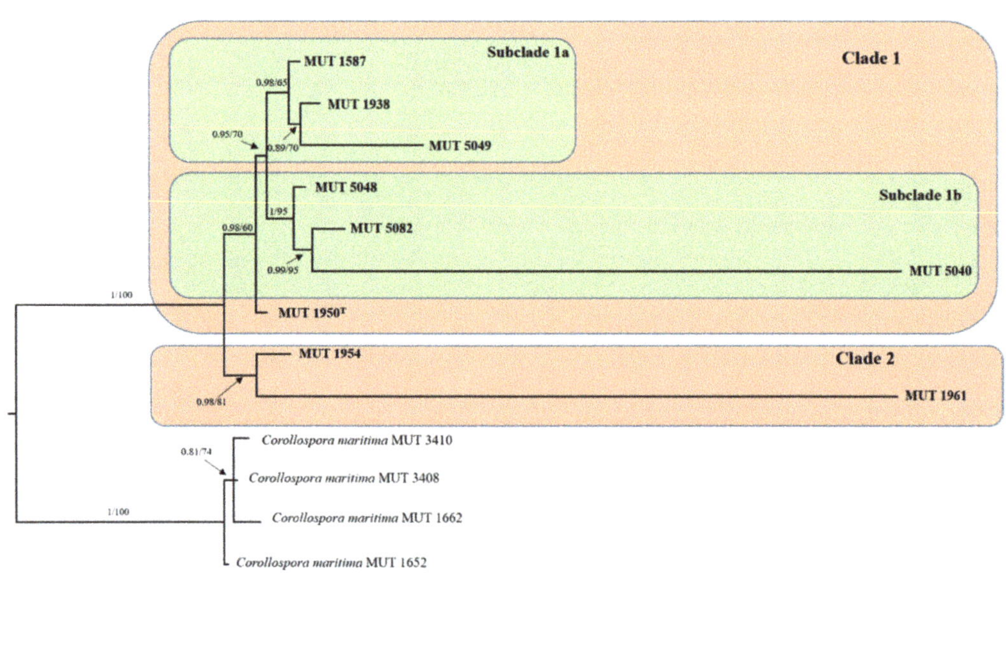

Figure 2. Phylogenetic inference based on a combined nrITS, nrSSU, nrLSU, RPB1, RPB2, TEF-1α, and βTUB dataset. The tree is midrooted. Branch numbers indicate BYPP and BS values; ^{T} Type Strain; Bar = expected changes per site (0.004).

3.2. Taxonomy

Corollospora mediterranea sp. nov. A. Poli, E. Bovio, G.C. Varese and V. Prigione, MYCOBANK: MB 839640, Type. Italy, Sicily, Mediterranean Sea, Gela (CL), July 2013, from seawater contaminated by an oil spill, R. Denaro, MUT 1950 holotype, living culture permanently-preserved in metabolically inactively state by deep-freezing at MUT. Additional material examined. Italy, Sicily, Mediterranean Sea, Gela (CL), July 2013, from seawater contaminated by an oil spill, R. Denaro MUT 1938, MUT 1954 and MUT 1961. Italy, Apulia, Mediterranean Sea, Porto Badisco (LE), July 2011, from driftwood, L. Garzoli, MUT 1587. Italy, Tuscany, Mediterranean Sea, Elba Island (LI), from the seagrass *Posidonia oceanica*, March 2010, R. Mussat-Sartor and N. Nurra, MUT 5040, MUT 5048, MUT 5049 and MUT 5082. Etymology. In reference to the Mediterranean Sea. Description. Growing actively on *Pinus pinaster* and *Quercus ruber* cork. *Hyphae* two types observed, one thin (1 μm wide) and hyaline, one thicker (3 μm wide) and melanized. *Chlamydospores* numerous, in chain globose to subglobose, 4–13 × 4–11 μm diameter (Figure 3). Sexual morph not observed. Asexual morph with differentiated conidiogenesis not observed. Colony description. Colonies on MEASW attaining 38–54 mm diam after 28 days at 21 °C (Figure 4), mycelium white, grey, grey-green, sometimes with pinkish shades; dense and feltrose, occasionally umbonate in the middle, with radial grooves; reverse from light brown to dark green, occasionally with concentric rings, lighter to the edges. Neither soluble pigments nor exudates were observed (Figure 3).

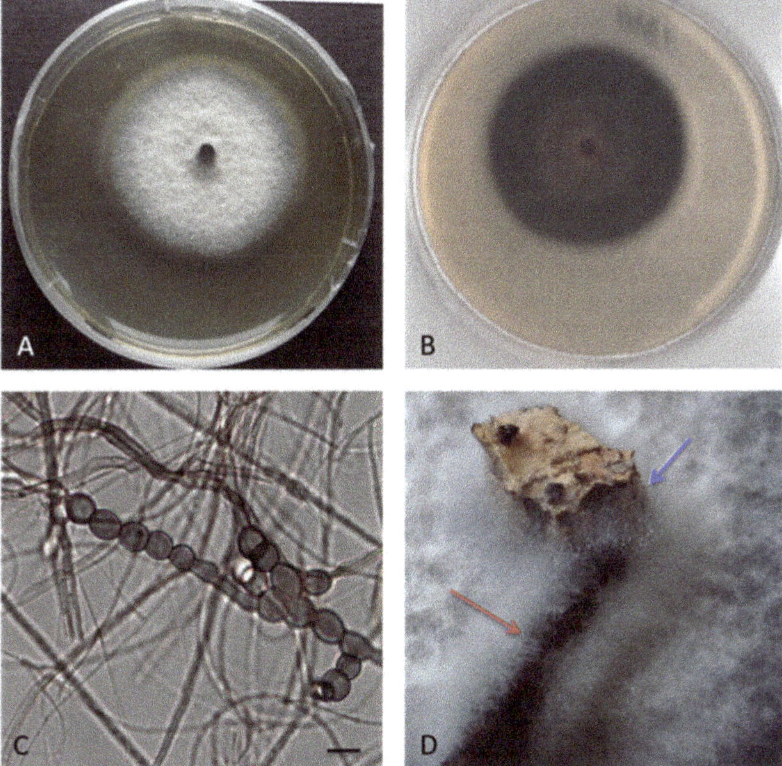

Figure 3. *Corollospora mediterranea* sp. nov. MUT 1950 (holotype), the 28-days-old colony at 21 °C on MEASW (**A**) and reverse (**B**); chlamydospores in the chain (**C**); mycelial growth on *Pinus pinaster* wood (red arrow) and *Quercus ruber* cork (blue arrow) (**D**). Scale bars: 10 μm.

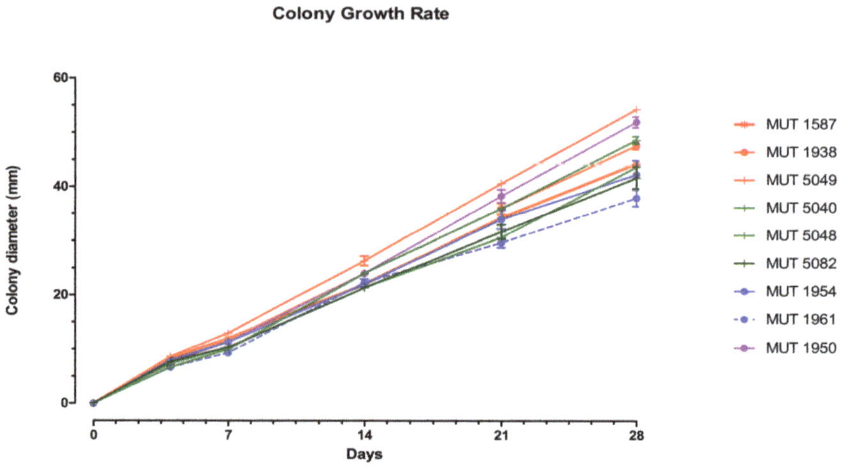

Colony Growth Rate

Figure 4. Growth rate on MEASW of the strains investigated over 28 days. The symbol indicates the isolation substrate (asterisk: driftwood; full circle: oil spill; vertical line: leaves of *Posidonia oceanica*); the color indicates the position in the phylogenetic tree (red: Clade 1, Subclade 1a; green: Clade 1, Subclade 1b; purple: Clade 1; blue: Clade 2).

4. Discussion

The description of the strains investigated in this study was particularly tricky and complicated since neither asexual nor sexual reproductive structures developed in axenic conditions. As a consequence, it was not possible to describe the range of diagnostic traits amongst these newly identified lineages.

To better characterize these fungi, we tried to mimic the saline environment by using a culture medium supplemented with seawater, since it is well known that only this method supports a measurable growth of vegetative mycelium [33]. The family Halospaheriaceae, to whom the genus *Corollospora* belongs, includes the largest number of marine lignicolous species. It was thus realistic to induce sporulation by placing wood and cork specimens on the colony surface prior to transfer them into seawater. Despite wood colonization, sporulation did not occur, whereas chlamydospores were abundantly produced. Likewise, strains of *Corollospora* sp. isolated from intertidal decayed wood of mangrove trees and driftwood in Saudi Arabia did not develop reproductive structures but only chlamydospores that were much bigger and not comparable to those observed in this study [30]. It must be considered that strictly vegetative growth with no sporulation is a common feature of marine fungi [46–48] that are likely to rely on their dispersion to hyphal fragments and/or chlamydospores. Most probably, a necessary requisite for fungi to develop reproductive structures is the occurrence of those environmental conditions these organisms are adapted to (e.g., high salinity, low temperature, high pressure, wet-dry cycles, etc.).

The phylogenetic analysis based on ribosomal genes (nrITS, nrLSU, nrSSU) and RPB1, shows a clear distance between the strains under investigation and the other species of *Corollospora*, *Corollospora maritima* being the closest one. This new and strongly supported clade may include one or more novel species. Hypothetically, the tree highlights the presence of three clusters that include: (i) MUT 1954 and MUT 1961; (ii) MUT 1938, MUT 5040, MUT 5049 and MUT 5082; (iii) MUT 1587, MUT 1950, and MUT 5048 (Figure 1). Bearing in mind the conclusion drawn by Campbell et al. [15] that indicated the need of sequencing more genetic markers to clarify the relations among the species of the genus *Corollospora*, a dataset focused on this new group, together with *C. maritima*, was built with the addition of three more protein-coding genes, namely RPB2, TEF-1α and β-TUB (Figure 2). From one side, this strongly supported tree points out once more the distance from *C. maritima*, strengthening the idea that we are dealing with a new species. It is undeniably complicated to define a fungal species, and *Corollospora mediterranea* was here established following the recommendations outlined by Jeewon and Hyde [49], who, dealing with the issue of the species boundaries and identification of new taxa, pointed out a number of key elements to follow. Notably, all the ITS sequences (including 5.8S) analyzed are longer than the minimum 450 base pairs required and display a percentage of identity with the closest relative *C. maritima* < 95%. In addition, as recommended, the strongly supported phylogenetic tree includes the minimum number (4–5) of closely related taxa of the same genus (Figure 1). A thorough inspection of the tree shown in Figure 2 reveals the presence of two clusters: the former (Clade 1) consists of MUT 1950 and two subclusters (MUT 5040, MUT 5048 and MUT 5082; MUT 1587, MUT 1938, MUT 5049), the latter (Clade 2) includes MUT 1954 and MUT 1961. On the other hand, the lack of distinct micro-morphological traits (neither sexual nor asexual morph with differentiated conidiogenesis detected) and the huge variability observed among the colonies (in terms of growth rate, texture, surface and reverse color) lead us to introduce the *C. mediterranea* species complex (CMSC), since we could not discern the species boundaries with certainty (Figures 5 and 6). Indeed, the term "species complex" comes to help taxonomists when: (i) a group of organisms may represent more than one species; (ii) morphological features are overlapping due to extreme variability; (iii) the species may be somehow related although no certain assumption can be assessed. Furthermore, no clear correlation between colony features, growth rates, source of isolation, and/or phylogenetic position was noticed. In general, researchers introduce a species complex when facing a problematic topic. This is the case for example of *Fusarium oxysporum* species complex (FOSC) [50,51], *Fusarium solani* species complex (FSSC) [52], *Wallemia sebi* species complex (WSSC) [53], *Colletotrichum acutatum* species complex (CASC) [54] and many others. Most of these cases were resolved with a revision of the species complex, where individual species were identified based on a multi-loci phylogeny. However, in our study, this approach did not lead to a sharp resolution of the complex. The genome sequencing of all the strains investigated would be an option to sort out this intriguing issue. This approach would possibly reveal those genetic regions that may allow an easy distinction of the species within the complex [55]. On the other hand, the same goal could be achieved by the analysis of secondary metabolites [53]. An investigation of this sort is important also from another point of view: it is now recognized that marine fungi are a reservoir of novel active metabolites that can be harnessed for pharmaceutical, nutraceutical, cosmetic, and environmental purposes.

Kirk and Gordon [10] demonstrated that a number of strains of *C. maritima*, *C. lacera*, and *C. intermedia* were capable of using hexadecane as the sole carbon source, and assumed that the genus *Corollospora* could find utility in petroleum degradation. Supporting this idea is the isolation of 4 strains of *C. mediterranea* (MUT 1938, MUT 1950, MUT 1954, and MUT 1961) from a site chronically contaminated by an oil spill in the Mediterranean Sea [28]. The aforementioned findings indicate the great versatility and adaptability of these fungi to hydrocarbons-contaminated environments, with the consequent possibility of being used as bioremediators. In addition, knowing that the species of *Corollospora* belong to a family of lignicolous fungi and that are rich in cellulase and lignin-degrading enzymes [7], their retrieval from leaves of the seagrass *P. oceanica* (MUT 5040, MUT 5048, MUT 5049, and MUT 5082) and from driftwood (MUT 1587) does not come as a surprise. It is, therefore, reasonable to assume a role of *C. mediterranea* in degrading and recycling organic matter, making them available for other organisms.

Figure 5. *Corollospora mediterranea* sp. nov. 28-days-old colonies at 21 °C on MEASW: (**A**) MUT 1587, (**B**) MUT 1938, (**C**) MUT 5049 (strains belonging to the Clade 1, Subclade 1a); (**D**) MUT 5048, (**E**) MUT 5082, (**F**) MUT 5040 (strains belonging to the Clade 1, Subclade 1b); (**G**) MUT 1954, (**H**) MUT 1961 (strains belonging to the Clade 2).

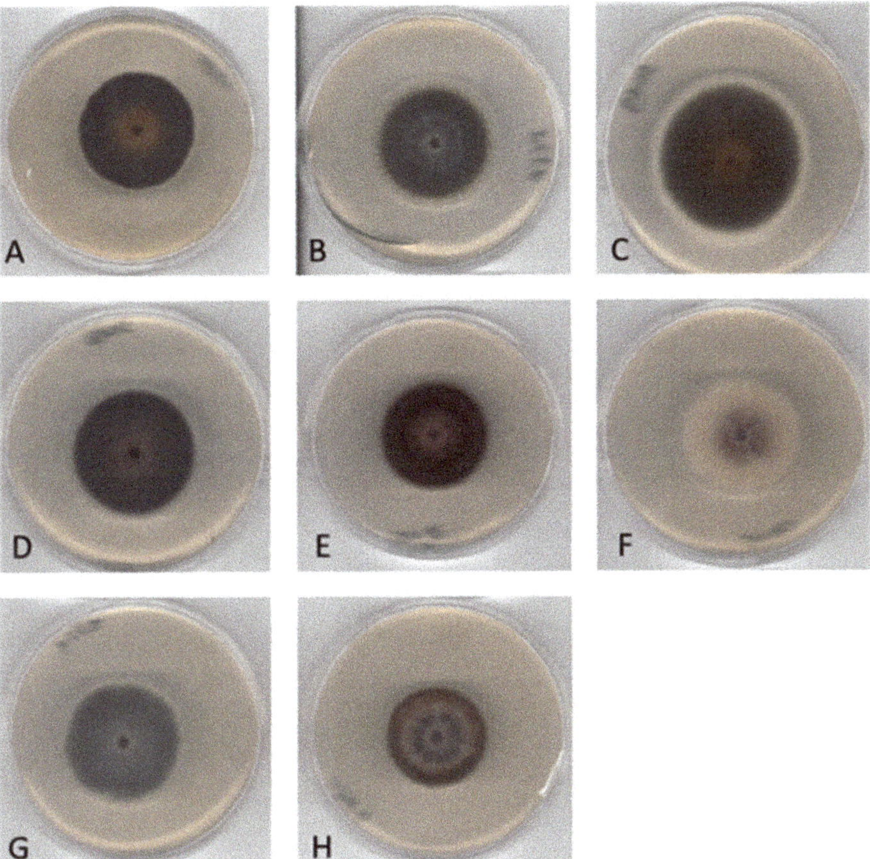

Figure 6. *Corollospora mediterranea* sp. nov. reverse of 28-days-old colonies at 21 °C on MEASW: (**A**) MUT 1587, (**B**) MUT 1938, (**C**) MUT 5049 (strains belonging to the Clade 1, Subclade 1a); (**D**) MUT 5048, (**E**) MUT 5082, (**F**) MUT 5040 (strains belonging to the Clade 1, Subclade 1b); (**G**) MUT 1954, (**H**) MUT 1961 (strains belonging to the Clade 2).

5. Conclusions

In conclusion, the retrieval of *C. mediterranea* from different substrates, localities and in different sampling campaigns indicates its constant presence in (at least) the Mediterranean Sea and points out how the marine environments is still largely uninvestigated from a mycological point of view. Therefore, it is more and more glaring that the Oceans are a huge reservoir of unidentified microorganisms with a valuable biotechnological potential not yet disclosed. In the next future, we would aim at resolving the species complex, by applying more approaches to induce sexual and/or asexual sporulation. In parallel, all the strains studied in this work, will be investigated for the production of new and powerful bioactive molecules.

Author Contributions: Conceptualization, A.P., E.B., and V.P.; methodology, A.P., E.B., and V.P.; software, A.P.: validation, A.P., I.P., and V.P.; formal analysis, A.P., E.B., and V.P.; investigation, A.P., E.B., and V.P.; resources, V.P. and G.C.V.; data curation, A.P., I.P., and V.P.; writing—original draft preparation, A.P. and V.P.; writing—review and editing, A.P., V.P., and G.C.V.; visualization, A.P. and V.P.; supervision, V.P. and G.C.V.; project administration, A.P., V.P., and G.C.V.; funding acquisition, G.C.V. All authors have read and agreed to the published version of the manuscript.

Funding: This research was funded by the University of Torino (ex 60%).

Institutional Review Board Statement: Not applicable.

Informed Consent Statement: Not applicable.

Data Availability Statement: Sequence data are available at Genbank (NCBI) under the accession numbers reported in the manuscript.

Acknowledgments: The authors would like to thank JRU MIRRI-IT for technical and scientific support.

Conflicts of Interest: The authors declare no conflict of interest.

References

1. Jones, E.B.G.; Pang, K.-L.; Abdel-Wahab, M.A.; Scholz, B.; Hyde, K.D.; Boekhout, T.; Ebel, R.; Rateb, M.E.; Henderson, L.; Sakayaroj, J.; et al. An online resource for marine fungi. *Fungal Divers.* **2019**, *96*, 347–433. [CrossRef]
2. Jones, E.G.; Ju, W.-T.; Lu, C.-L.; Guo, S.-Y.; Pang, K.-L. The halosphaeriaceae revisited. *Bot. Mar.* **2017**, *60*, 453–468. [CrossRef]
3. Abdel-Wahab, M.A.; Nagahama, T.; Abdel-Aziz, F.A. Two new Corollospora species and one new anamorph based on morphological and molecular data. *Mycoscience* **2009**, *50*, 147–155. [CrossRef]
4. Hsieh, S.Y.; Moss, S.T.; Jones, E.B.G. Ascoma development in the marine ascomycete *Corollospora gracilis* (Halosphaeriales, Hypocreomycetidae, Sordariomycetes). *Bot. Mar.* **2007**, *50*, 302–313. [CrossRef]
5. Walker, A.K.; Robicheau, B.M. Fungal diversity and community structure from coastal and barrier island beaches in the United States Gulf of Mexico. *Sci. Rep.* **2021**, *11*, 3889. [CrossRef] [PubMed]
6. Velez, P.; Gasca-Pineda, J.; Nakagiri, A.; Hanlin, R.T.; Gonzalez, M.C. Genetic diversity and population structure of *Corollospora maritima* sensu lato: New insights from population genetics. *Bot. Mar.* **2016**, *59*, 307–320. [CrossRef]
7. Raghukumar, S. *Fungi in Coastal and Oceanic Marine Ecosystems: Marine Fungi*; Springer: Cham, Switzerland, 2017.
8. Overy, D.P.; Bayman, P.; Kerr, R.G.; Bills, G.F. An assessment of natural product discovery from marine (sensu strictu) and marine-derived fungi. *Mycology* **2014**, *5*, 145–167. [CrossRef]
9. Overy, D.P.; Rama, T.; Oosterhuis, R.; Walker, A.K.; Pang, K.L. The Neglected Marine Fungi, Sensu stricto, and Their Isolation for Natural Products' Discovery. *Mar. Drugs* **2019**, *17*, 42. [CrossRef]
10. Kirk, P.W.; Gordon, A.S. Hydrocarbon degradation by fiilamentous marine higher fungi. *Mycologia* **1988**, *80*, 776–782. [CrossRef]
11. Liberra, K.; Jansen, R.; Lindequist, U. Corollosporine, a new phthalide derivative from the marine fungus *Corollospora maritima* Werderm. 1069. *Pharmazie* **1998**, *53*, 578–581. [CrossRef]
12. Furuya, K.; Okudaira, M.; Shindo, T.; Sato, A. *Corollospora pulchella*, a marine fungus producing antibiotics, melinacidins III, IV and gancidin W. *Annu. Rep. Sankyo Res. Lab.* **1985**, *37*, 140–142.
13. Alvi, K.A.; Casey, A.; Nair, B.G. Pulchellalactam: A CD45 protein tyrosine phosphatase inhibitor from the marine fungus *Corollospora pulchella*. *J. Antibiot.* **1998**, *51*, 515–517. [CrossRef]
14. MacKenzie, S.E.; Gurusamy, G.S.; Piorko, A.; Strongman, D.B.; Hu, T.M.; Wright, J.L.C. Isolation of sterols from the marine fungus *Corollospora lacera*. *Can. J. Microbiol.* **2004**, *50*, 1069–1072. [CrossRef] [PubMed]
15. Campbell, J.; Shearer, C.A.; Mitchell, J.I.; Eaton, R.A. Corollospora revisited: A molecular approach. In *Fungi in Marine Environment*; Hyde, K.D., Ed.; (Fungal Diversity Research Series); Fungal Diversity Press: Hong Kong, China, 2002; Volume 7, pp. 15–33.
16. Jones, E.G.; Johnson, R.; Moss, S. Taxonomic studies of the Halosphaeriaceae: Corollospora Werdermann. *Bot. J. Linn. Soc.* **1983**, *87*, 193–212. [CrossRef]
17. Lutley, M.; Wilson, I.M. Development and fine structure of ascospores in the marine fungus *Ceriosporopsis halima*. *Trans. Br. Mycol. Soc.* **1972**, *58*, 393–402. [CrossRef]
18. Lutley, M.; Wilson, I.M. Observations on the fine structure of ascospores of marine fungi: Halosphaeria appendiculata, *Torpedospora radiata* and *Corollospora maritima*. *Trans. Br. Mycol. Soc.* **1972**, *59*, 219–227. [CrossRef]
19. Jones, E.B.G.; Sakayaroj, J.; Suetrong, S.; Somrithipol, S.; Pang, K.L. Classification of marine Ascomycota, anamorphic taxa and Basidiomycota. *Fungal Divers.* **2009**, *35*, 187.
20. Koch, J. Some Lignicolous Marine Fungi From Thailand, including 2 New Species. *Nord. J. Bot.* **1986**, *6*, 497–499. [CrossRef]
21. Kohlmeyer, J.; Volkmannkohlmeyer, B. Corollospora armoricana sp. nov An Arenicolous Ascomycete from Brittany (France). *Can. J. Bot. Rev. Can. Bot.* **1989**, *67*, 1281–1284. [CrossRef]
22. McKeown, T.A.; Moss, S.T.; Jones, E.B.G. Ultrastructure of the melanized ascospores of *Corollospora fusca*. *Can. J. Bot. Rev. Can. Bot.* **1996**, *74*, 60–66. [CrossRef]
23. Jones, E.B.G. Fifty years of marine mycology. *Fungal Divers.* **2011**, *50*, 73–112. [CrossRef]
24. Sundari, R.; Vikineswary, S.; Yusoff, M.; Jones, E.B.G. Corollospora besarispora, a new arenicolous marine fungus from Malaysia. *Mycol. Res.* **1996**, *100*, 1259–1262. [CrossRef]
25. Schoch, C.L.; Crous, P.W.; Groenewald, J.Z.; Boehm, E.W.A.; Burgess, T.I.; de Gruyter, J.; de Hoog, G.S.; Dixon, L.J.; Grube, M.; Gueidan, C.; et al. A class-wide phylogenetic assessment of Dothideomycetes. *Stud. Mycol.* **2009**, *64*, 1–15. [CrossRef]

26. Schoch, C.L.; Seifert, K.A.; Huhndorf, S.; Robert, V.; Spouge, J.L.; Levesque, C.A.; Chen, W.; Bolchacova, E.; Voigt, K.; Crous, P.W.; et al. Nuclear ribosomal internal transcribed spacer (ITS) region as a universal DNA barcode marker for Fungi. *Proc. Natl. Acad. Sci. USA* **2012**, *109*, 6241–6246. [CrossRef] [PubMed]
27. Panno, L.; Bruno, M.; Voyron, S.; Anastasi, A.; Gnavi, G.; Miserere, L.; Varese, G.C. Diversity, ecological role and potential biotechnological applications of marine fungi associated to the seagrass *Posidonia oceanica*. *New Biotechnol.* **2013**, *30*, 685–694. [CrossRef] [PubMed]
28. Bovio, E.; Gnavi, G.; Prigione, V.; Spina, F.; Denaro, R.; Yakimov, M.; Calogero, R.; Crisafi, F.; Varese, G.C. The culturable mycobiota of a Mediterranean marine site after an oil spill: Isolation, identification and potential application in bioremediation. *Sci. Total Environ.* **2017**, *576*, 310–318. [CrossRef] [PubMed]
29. Garzoli, L.; Gnavi, G.; Tamma, F.; Tosi, S.; Varese, G.C.; Picco, A.M. Sink or swim: Updated knowledge on marine fungi associated with wood substrates in the Mediterranean Sea and hints about their potential to remediate hydrocarbons. *Prog. Oceanogr.* **2015**, *137*, 140–148. [CrossRef]
30. Abdel-Wahab, M.A.; Hodhod, M.S.; Bahkali, A.H.A.; Jones, E.B.G. Marine fungi of Saudi Arabia. *Bot. Mar.* **2014**, *57*, 323–335. [CrossRef]
31. Dayarathne, M.C.; Wanasinghe, D.N.; Devadatha, B.; Abeywickrama, P.; Jones, E.B.G.; Chomnunti, P.; Sarma, V.V.; Hyde, K.D.; Lumyong, S.; McKenzie, E.H.C. Modern taxonomic approaches to identifying diatrypaceous fungi from marine habitats, with a novel genus Halocryptovalsa Dayarathne & KDHyde, gen. nov. *Cryptogam. Mycol.* **2020**, *41*, 21–67. [CrossRef]
32. Jones, E.G.; Pang, K.-L. (Eds.) *Marine Fungi and Fungal-Like Organisms*; Walter de Gruyter: Berlin, Germany, 2012.
33. Poli, A.; Bovio, E.; Ranieri, L.; Varese, G.C.; Prigione, V. News from the Sea: A New Genus and Seven New Species in the Pleosporalean Families Roussoellaceae and Thyridariaceae. *Diversity* **2020**, *12*, 144. [CrossRef]
34. Panebianco, C.; Tam, W.Y.; Jones, E.B.G. The effect of pre-inoculation of balsa wood by selected marine fungi and their effect on subsequent colonisation in the sea. *Fungal Divers.* **2002**, *10*, 77–88.
35. White, T.J.; Bruns, T.; Lee, S.; Taylor, J. Amplification and direct sequencing of fungal ribosomal RNA genes for phylogenetics. *PCR Protoc. Guide Methods Appl.* **1990**, *18*, 315–322.
36. Vilgalys, R.; Hester, M. Rapid genetic identification and mapping of enzymatically amplified ribosomal DNA from several *Cryptococcus species*. *J. Bacteriol.* **1990**, *172*, 4238–4246. [CrossRef] [PubMed]
37. Matheny, P.B.; Wang, Z.; Binder, M.; Curtis, J.M.; Lim, Y.W.; Nilsson, R.H.; Hughes, K.W.; Hofstetter, V.; Ammirati, J.F.; Schoch, C.L.; et al. Contributions of rpb2 and tef1 to the phylogeny of mushrooms and allies (Basidiomycota, Fungi). *Mol. Phylogenetics Evol.* **2007**, *43*, 430–451. [CrossRef]
38. Glass, N.L.; Donaldson, G.C. Development of primer sets designed for use with the PCR to amplify conserved genes from filamentous ascomycetes. *Appl. Environ. Microbiol.* **1995**, *61*, 1323–1330. [CrossRef]
39. Raja, H.A.; Baker, T.R.; Little, J.G.; Oberlies, N.H. DNA barcoding for identification of consumer-relevant mushrooms: A partial solution for product certification? *Food Chem.* **2017**, *214*, 383–392. [CrossRef]
40. Liu, Y.J.; Whelen, S.; Hall, B.D. Phylogenetic relationships among ascomycetes: Evidence from an RNA polymerse II subunit. *Mol. Biol. Evol.* **1999**, *16*, 1799–1808. [CrossRef]
41. Sakayaroj, J.; Pang, K.L.; Jones, E.B.G. Multi-gene phylogeny of the Halosphaeriaceae: Its ordinal status, relationships between genera and morphological character evolution. *Fungal Divers.* **2011**, *46*, 87–109. [CrossRef]
42. Vaidya, G.; Lohman, D.J.; Meier, R. SequenceMatrix: Concatenation software for the fast assembly of multi-gene datasets with character set and codon information. *Cladistics* **2011**, *27*, 171–180. [CrossRef]
43. Darriba, D.; Taboada, G.L.; Doallo, R.; Posada, D. jModelTest 2: More models, new heuristics and parallel computing. *Nat. Methods* **2012**, *9*, 772. [CrossRef] [PubMed]
44. Stamatakis, A. RAxML version 8: A tool for phylogenetic analysis and post-analysis of large phylogenies. *Bioinformatics* **2014**, *30*, 1312–1313. [CrossRef]
45. Ronquist, F.; Teslenko, M.; van der Mark, P.; Ayres, D.L.; Darling, A.; Hohna, S.; Larget, B.; Liu, L.; Suchard, M.A.; Huelsenbeck, J.P. MrBayes 3.2: Efficient Bayesian Phylogenetic Inference and Model Choice Across a Large Model Space. *Syst. Biol.* **2012**, *61*, 539–542. [CrossRef]
46. Gnavi, G.; Garzoli, L.; Polil, A.; Prigione, V.; Burgaud, G.; Varese, G.C. The culturable mycobiota of Flabellia petiolata: First survey of marine fungi associated to a Mediterranean green alga. *PLoS ONE* **2017**, *12*. [CrossRef] [PubMed]
47. Garzoli, L.; Poli, A.; Prigione, V.; Gnavi, G.; Varese, G.C. Peacock's tail with a fungal cocktail: First assessment of the mycobiota associated with the brown alga Padina pavonica. *Fungal Ecol.* **2018**, *35*, 87–97. [CrossRef]
48. Poli, A.; Vizzini, A.; Prigione, V.; Varese, G.C. Basidiomycota isolated from the Mediterranean Sea—Phylogeny and putative ecological roles. *Fungal Ecol.* **2018**, *36*, 51–62. [CrossRef]
49. Jeewon, R.; Hyde, K.D. Establishing species boundaries and new taxa among fungi: Recommendations to resolve taxonomic ambiguities. *Mycosphere* **2016**, *7*, 1669–1677. [CrossRef]
50. Kistler, H.C. Genetic diversity in the plant-pathogenic fungus *Fusarium oxysporum*. *Phytopathology* **1997**, *87*, 474–479. [CrossRef]
51. O'Donnell, K.; Cigelnik, E. Two divergent intragenomic rDNA ITS2 types within a monophyletic lineage of the fungus Fusarium are nonorthologous. *Mol. Phylogenetics Evol.* **1997**, *7*, 103–116. [CrossRef]
52. O'Donnell, K. Molecular phylogeny of the Nectria haematococca-Fusarium solani species complex. *Mycologia* **2000**, *92*, 919–938. [CrossRef]

53. Jancic, S.; Nguyen, H.D.T.; Frisvad, J.C.; Zalar, P.; Schroers, H.J.; Seifert, K.A.; Gunde-Cimerman, N. Taxonomic Revision of the Wallemia sebi Species Complex. *PLoS ONE* **2015**, *10*, e0125933. [CrossRef]
54. Damm, U.; Cannon, P.F.; Woudenberg, J.H.C.; Crous, P.W. The *Colletotrichum acutatum* species complex. *Stud. Mycol.* **2012**, *73*, 37–113. [CrossRef] [PubMed]
55. Matute, D.R.; Sepulveda, V.E. Fungal species boundaries in the genomics era. *Fungal Genet. Biol.* **2019**, *131*, 9. [CrossRef] [PubMed]

Article

Extra-Heavy Crude Oil Degradation by *Alternaria* sp. Isolated from Deep-Sea Sediments of the Gulf of Mexico

Lucia Romero-Hernández [1], Patricia Velez [2], Itandehui Betanzo-Gutiérrez [3], María Dolores Camacho-López [1], Rafael Vázquez-Duhalt [3] and Meritxell Riquelme [1,*]

1 Departamento de Microbiología, Centro de Investigación Científica y de Educación Superior de Ensenada (CICESE), Ensenada, Baja California 22860, Mexico; lromero@cicese.edu.mx (L.R.-H.); mcamacho@cicese.edu.mx (M.D.C.-L.)
2 Departamento de Botánica, Instituto de Biología, Universidad Nacional Autónoma de México, Ciudad de México 04510, Mexico; pvelez@ib.unam.mx
3 Centro de Nanociencias y Nanotecnología (CNyN), Universidad Nacional Autónoma de México (UNAM), Baja California 22860, Mexico; ibetanzo@cigom.org (I.B.-G.); rvd@cnyn.unam.mx (R.V.-D.)
* Correspondence: riquelme@cicese.mx

Abstract: The Gulf of Mexico (GoM) is an important source of oil for the United States and Mexico. There has been growing interest, particularly after the Deepwater Horizon oil spill, in characterizing the fungal diversity of the GoM and identifying isolates for use in the bioremediation of petroleum in the event of another spill. Most studies have focused on light crude oil bioremediation processes, while heavy crude oil (HCO) and extra-heavy crude oil (EHCO) have been largely ignored. In this work, we evaluated the ability of fungal isolates obtained from deep-sea sediments of the Mexican economic exclusive zone (EEZ) of the GoM to degrade HCO (16–20° API) and EHCO (7–10° API). *Alternaria* sp., *Penicillium* spp., and *Stemphylium* sp. grew with HCO as the sole carbon source. Remarkably, *Alternaria* sp. was the only isolate able to grow with EHCO as the sole carbon source, degrading up to 25.6% of the total EHCO and 91.3% of the aromatic fraction, as demonstrated by gas chromatography analysis of the saturate, aromatic, and polar fractions. These findings proved to be significant, identifying *Alternaria* sp. as one of the few fungi reported so far capable of degrading untreated EHCO and as a suitable candidate for bioremediation of EHCO in future studies.

Keywords: heavy crude oil; mycodegradation; deep-sea fungi; bioremediation; fungal isolation

Citation: Romero-Hernández, L.; Velez, P.; Betanzo-Gutiérrez, I.; Camacho-López, M.D.; Vázquez-Duhalt, R.; Riquelme, M. Extra-Heavy Crude Oil Degradation by *Alternaria* sp. Isolated from Deep-Sea Sediments of the Gulf of Mexico. *Appl. Sci.* **2021**, *11*, 6090. https://doi.org/10.3390/app11136090

Academic Editors: Anna Poli and Valeria Prigione

Received: 20 May 2021
Accepted: 23 June 2021
Published: 30 June 2021

Publisher's Note: MDPI stays neutral with regard to jurisdictional claims in published maps and institutional affiliations.

1. Introduction

Crude oil has become the most important source of energy for humankind. It has been estimated that, by 2035, it will supply more than one-third of the total global energy demands [1,2]. Light crude oil has been the primary option to produce refined products for everyday use, such as gasoline, diesel, and kerosene, among many others [3,4]. However, reserves of conventional light crude oil are rapidly being depleted all around the world, while, at the same time, the demand for fossil fuel is continually increasing as a result of human population growth and economic development [5]. Alternative green energy sources with lower carbon footprints (e.g., solar and eolic) have improved to become economically viable, but they are nonetheless insufficient to supply the increasing energy demand [6]. Therefore, the exploitation of unconventional petroleum fuel resources (tar sands and extra-heavy oil deposits) might emerge as an alternative source of fossil fuel in the next decades [7]. Heavy and extra-heavy crude oil (HCO and EHCO, respectively) make up 70% of the world's oil reserves [7]. However, their high content in resins and asphaltenes represents a significant challenge for its extraction, production, and refinement, in addition to the associated risk of accidental spills. Additionally, polycyclic aromatic hydrocarbons (PHAs) are also present in a significant ratio in HCO and EHCO. These fractions are one of the major concerns in oil spills, given their resistance to microbial degradation [8], long-term persistence in the environment [9], and ecotoxicity [10–12].

Bioremediation is a well-documented option that involves applying microorganisms to neutralize petroleum in the environment [13]. It targets those microbes capable of synthesizing enzymes that cleave C:H bounds from petroleum hydrocarbons and transforms them into harmless and less persistent molecules [14]. However, while the microbial biodegradation of hydrocarbons and crude oils has been widely studied for decades, most studies have focused on light crude oils and their low molecular weight fractions or purified components [15]. In contrast, the microbial biodegradation of HCO and EHCO, particularly by fungi, has been scarcely studied.

Only a few fungal strains isolated from extreme environments are able to degrade asphaltenes and high molecular weight PAHs [16]. *Aspergillus fischeri* (previously *Neosartorya fischeri*), isolated from an asphalt lake in Venezuela, was the first microorganism able to grow using purified asphaltenes as the sole carbon source [17]. This fungus set a unique precedent demonstrating its remarkable ability to degrade extremely complex hydrocarbon molecules, especially high molecular weight PHAs [18]. The extremophilic fungus *Pestalotiopsis palmarum* BM-04 isolated from an asphalt lake in Venezuela synthesizes oxidative exoenzymes that catalyze the biotransformation of maltenes, asphaltenes, and the petroporphyrins-rich fraction of biotreated EHCO [19]. *Paecilomyces variotii*, *Fusarium decemcellulare*, *Candida palmioleophila*, and *Pichia guilliermondii*, isolated from a petroleum activity site in Indonesia, were able to degrade 10–15% of resins and asphaltenes using BAL150 light crude oil as the sole carbon source [20]. A chloroperoxidase obtained from *Caldariomyces fumago* transformed the porphyrin-free asphaltene fraction recovered from Maya HCO, reducing it by 24% [21]. Furthermore, *Daedaleopsis* sp., isolated from a forest in Iran, showed biodegradation of 38% of the asphaltene and aromatic fractions of HCO [22].

The Gulf of Mexico (GoM) is rich in petroleum reservoirs, and since 2017, their ultra-deep waters have been exploited to obtain crude oil [23]. Therefore, it represents an ideal prospecting source to isolate indigenous fungal strains adapted to deep-sea conditions and capable of degrading crude oil. In a previous study, the fungal taxa *Aureobasidium* sp., *Penicillium brevicompactum*, *Penicillium* sp., *Phialocephala* sp., and *Cladosporium* sp. isolated from sediments of the Mexican economic exclusive zone (EEZ) of the GoM were identified as able to metabolize alkane and alkene, long-chain hydrocarbons poorly degraded by bacteria, as the sole carbon sources [24]. Furthermore, the RNA-seq expression analysis of *Penicillium* sp. cultured with hexadecane or 1-hexadecene showed the upregulation of genes involved in transmembrane transport, metabolism of six-carbon carbohydrates, and nitric oxide pathways [24].

In this study, we set out to identify culturable strains of marine fungi native to the seafloor of the Mexican EEZ of the GoM with the ability to degrade untreated EHCO. *Alternaria* sp. was the only isolate identified as capable of growing using EHCO as the sole carbon source and degrading a significant portion of the whole EHCO.

2. Materials and Methods

2.1. Sampling

The deep-sea sediment samples were collected from twenty-four stations located in Coatzacoalcos (also known as the Salina basin) and in the Perdido fold belt in the Mexican EEZ of the GoM (Figure 1). Samples were collected employing a Reineck box corer at different depths (Table S1) during the Metagenomics (MET-II) campaign that took place in 2017 onboard the research vessel Justo Sierra (UNAM). Subsamples were taken from the first 20 cm of the sediment collected with the box corer using pre-cut 50-mL sterile syringes. Syringes were sealed with plastic film and stored at 4 °C and transported to the laboratory for processing.

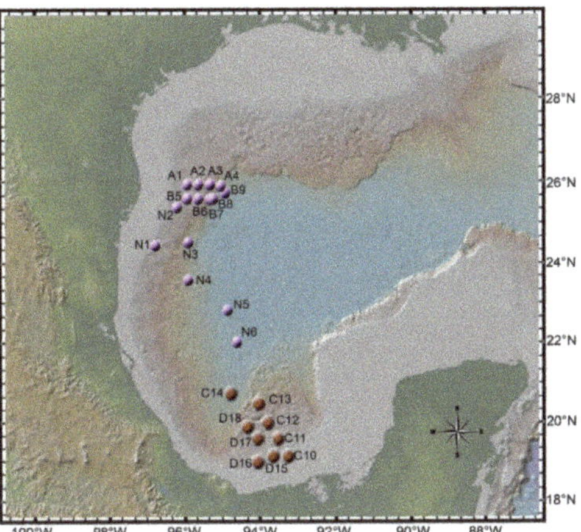

Figure 1. Localization of twenty-four sampled stations in the Gulf of Mexico, nine in Coatzacoalcos, also known as the Salina basin (orange dots), and fifteen in the Perdido fold belt region (purple dots).

2.2. Fungal Isolation and Identification

Sediment samples were processed as described previously [24]. Briefly, fungi were initially isolated in a five-fold diluted medium. For maintenance, fungal isolates were grown on a potato dextrose agar (PDA) medium. Colony morphology and pigmentation were registered using a Nikon D3100 Digital Camera (Nikon Inc., Tokyo, Japan). In addition, to analyze the microscopic characteristics, including the hyphal shape, morphology, size of spores, and conidiogenous cells, fungi were grown on water agar (1.5% agar). Samples were processed by using the inverted block technique [25]. Imaging was carried out by DIC (Differential Interference Contrast) on a Nikon eclipse Ti-E microscope with an oil 60× objective. Macroscopic and microscopic characteristics, as well as Taxonomic Keys [26] and the data available in MycoBank (http://www.mycobank.org/) (accessed on 9 December 2020), were considered to reach genus-level identification. Total genomic DNA was extracted from axenic isolates with the DNeasy Plant Mini Kit (Qiagen). Sequence-based taxonomic verification was achieved through phylogenetic analyses [27] using: (1) the internal transcribed spacer (ITS) region of the nuclear ribosomal DNA ITS1-5.8 rDNA using the primer set ITS1F (5-CTTGGTCATTTAGAGGAAGTAA-3) [28] and ITS4 (5-TCCTCCGCTTATTGATATGC-3) [29], (2) the 18S rRNA gene using primer set NS1 (5-GTAGTCATATGCTTGTCTC-3) and NS4 (5-CTTCCGTCAATTCCTTTAAG-3) [30], and (3) the β-tubulin gene (benA) with primer set Bt2A (5-GGTAACCAAATCGGTGCTGCTTTC-3) and Bt2b (5-ACCCTCAGTGTAGTGACCCTTGGC-3) [31] and with degenerate primers LR1 (5-RRCRACAARTTCGTGCCCCG-3) and LR2 (5-AGTGAACTGGTCACCYACAC-3), newly designed to amplify another region of the β-tubulin gene downstream of the region amplified with primer set Bt2A and Bt2b for cases in which that primer set did not work. The PCR products were purified using a QIAquick Gel Extraction Kit and commercially sequenced by Eton Bioscience, Inc., San Diego, CA, USA.

Consensus sequences were generated using Consed v29.0 [32–34] and compared to the GenBank Database through a BLAST search (http://www.ncbi.nlm.nih.gov/BLAST) (accessed on 9 December 2020) using the MegaBLAST algorithm to retrieve the reference sequences for the phylogenetic analyses. Only hit sequences with a minimum coverage of 98% were considered for the analyses, pondering accessions associated with vouchers and type material (Table S2). Uncultured/environmental sample sequences were not considered for the analyses. Sequences were aligned with the software UGENE v36.1 [35] using MUSCLE [36]. Phylogenetic trees were inferred by MEGA X [37,38] with the Maximum Likelihood (ML) algorithm, a bootstrap test of 1000 replications, and the K2+I (ITS and 18S) and K2+G and HKY+G (β-tubulin benA and LR1-LR2, respectively) models of nucleotide substitution [39]. Initial trees for the heuristic search were obtained automatically by applying Neighbor-Joining and BioNJ algorithms to a matrix of pairwise distances estimated using the Maximum Composite Likelihood (MCL) approach and then selecting the topology with a superior log likelihood value. The sequences are available in GenBank under the accession numbers MW412481-MW412485 for the 18S, MW412490-MW412494 for the ITS, and MZ392424-MZ392428 for the β-tubulin sequences (Table S3).

2.3. Growth with Crude Oil as a Sole Source of Carbon

Fungal conidia were cultured using crude oil as a sole carbon source to evaluate if oil metabolization occurred. Fungal strains were grown on PDA plates and incubated for five days at 30 °C to obtain conidia. Even though the strains could grow at room temperature, at 30 °C, an accelerated growth was observed. Conidia were recovered from the PDA plates using distilled sterile water, separated from mycelium through filtration, and washed twice with sterile distilled water. Conidia were counted using a Neubauer Chamber. Flasks containing 50 mL of modified Czapek minimal medium without a carbon source (4-g $NaNO_3$, 2-g K_2HPO_4, 1-g $MgSO_4 \bullet 7H_2O$, 1-g KCl, 0.02-g $FeSO_4 \bullet 7H_2O$, 32-g sea salt, and 0.5 mL of trace mineral solution) [40] were inoculated with 2×10^7 conidia. The flasks were supplemented with 0.5% of three different types of petroleum according to the American Petroleum Institute (API) classification: light crude oil (40° API from the well Xux in Tabasco, Mexico), heavy (16–20° API from the field Bacal in Tabasco), or extra-heavy (7–10° API from the well Ayatsil-Telek in Campeche, Mexico) crude oil as a sole source of carbon. The cultures were incubated at 30 °C under 150 rpm up to a month with weekly revisions.

2.4. Microscopic Analysis

To assess the ability of conidia to germinate using crude oil as a unique carbon source, hyphal development was observed by Differential Interference Contrast (DIC) microscopy using a Nikon Inverted Microscope Eclipse Ti-E with a 40 × 0.95 objective. Mycelial samples from each culture were taken from the petroleum–culture medium interface with a sterile bacteriological loop. The sample was spread onto a microscope slide with a drop of water.

Since *Alternaria* sp. was the only strain that displayed visible growth on the EHCO added to the modified Czapek medium, the samples were analyzed by Scanning Electron Microscopy (SEM) to image the mycelium–petroleum interface further. After three weeks, when mycelium was visible, the medium was poured off. A small portion of the mix containing petroleum and fungal biomass was fixed for two hours with 50 mL of a solution containing paraformaldehyde 1%, glutaraldehyde 2%, and potassium buffer 0.1 M (K_2HPO_4 1M and KH_2PO_4 1M, pH 6.8). The solution was removed by decantation, and the sample was covered with 2% osmium tetroxide and incubated for one hour. Finally, alcohol at different concentrations (25%, 50%, 75%, and 100%) was employed to fix and dehydrate the sample, which was imaged on a Hitachi SU3500 SEM.

2.5. Fungal Growth Quantification

To corroborate the mycelial growth of *Alternaria* sp. with EHCO as a unique carbon source, the protein concentration was quantified. The fungal cultures were incubated for a month at 30 °C under 150 rpm. Next, 20 mL of toluene were added to separate the mycelium from the EHCO. The culture was shaken vigorously and transferred to a centrifuge tube. The mixture was centrifuged at 15,000 rpm for 10 min, and the organic upper layer containing the remaining oil was removed. The pellet containing the fungal biomass was recovered by filtration with a vacuum pump through a Whatman Grade 1 Qualitative Filter Paper 1 0.25 mm. The mycelium was ground in a mortar with liquid nitrogen; mixed with 1 mL of lysis buffer (1-M Tris pH 7.4, 2-M KCl, 1-M $MgCl_2$, Triton X100, and one tablet of Complete Mini Protease Inhibitor Cocktail, Roche, CDMX, Mexico); and maintained for 30 min at 4 °C. After centrifugation at 12,000× *g* for 10 min, the supernatant was collected to measure the total protein content of the lysate. Abiotic cultures (EHCO and modified Czapek minimal medium) and cultures without EHCO were used as controls, and all experiments were performed in independent triplicates. An independent samples *t*-test was conducted to compare the mg/L of protein in control and treatment cultures. A Bio-Rad Protein Assay was used to quantify the protein content, and a BSA standard curve was included.

When all the strains were tested in the early stages of this project, *Penicillium* spp. were grown under different concentrations of light crude oil (0.25, 0.5%, and 0.75% *v/v*) as the sole carbon source to determine at which concentration they produced more biomass. It was unnecessary to add toluene to separate the mycelium from the oil, because there was no strong attachment between them, and the mycelium was easily recovered through filtration. Abiotic cultures (light crude oil and modified Czapek minimal medium) and cultures without light crude oil were used as controls, and all experiments were performed in independent triplicates. A one-way between treatments ANOVA was conducted to compare the effect of petroleum at different concentrations. Post hoc comparisons were conducted using Turkey's honestly significant difference (HSD).

2.6. Liquid–Liquid Extraction

Flasks containing 50 mL of modified Czapek medium supplemented with 0.5% *w/v* of EHCO (7–10° API) as unique carbon sources and inoculated with 2×10^7 conidia of *Alternaria* sp. were incubated at 30 °C and stirred at 150 rpm for a month. The remaining oil was extracted to perform the gas chromatography analysis. The medium was acidified with HCl and extracted with 10 mL of dichloromethane three times. The organic extract was recovered, dehydrated on an anhydrous Na_2SO_4 column, and dried on a rotary evaporator under 0.79 atm.

2.7. SAP Analysis

To analyze the percentage of degradation of the different fractions of EHCO by *Alternaria* sp., the crude oil was fractionated by a modified SARA (saturates, aromatics, resins, and asphaltenes) standard method (ASTM D2007). The modification consisted in obtaining the resins and asphaltenes components together in one fraction only. As a result, the saturates, aromatics, and polar (resins and asphaltenes) fractions were obtained (SAP). All the fractions were analyzed by gas chromatography. The solvents used *were* n-hexane to separate the saturates, toluene for the aromatics, and a mixture of methanol and dichloromethane (50:50) for resins and asphaltenes. The separation was made through a silica column, and three fractions were obtained according to the increasing polarity of the solvents. The fractions were evaporated on a rotary evaporator and analyzed on an Agilent 7820A Gas Chromatographer with a Zebron Inferno 20 m × 0.18 mm × 0.18 µm column (Phenomenex, Los Angeles, CA, USA), according to the EPA-8015 method (US EPA). The Zebron Inferno column, with a high temperature range, allowed to analyze the volatile asphaltenes. The total petroleum hydrocarbons (TPH) were quantified as the response of the flame ionization detector (FID). The GC program started at 50 °C (2 min), followed by a temperature ramp of 10 °C/min at 375 °C (5 min). The spitless injector was set up at 360 °C and the detector at 400 °C. Helium was used as carrier gas at a constant flow of 0.9 mL/min.

3. Results

3.1. All Fungal Isolates Recovered belonged to the Ascomycota

Twenty-seven fungal isolates were obtained from the twenty-four sites of the GoM. These were clustered into five morphotypes based on macroscopic and microscopic morphological features (Figure 2). Based on the phylogenetic analysis of the sequenced ITS region, 18S rRNA gene, and β-tubulin gene, we unequivocally identified *Cladosporium halotolerans* to the species level (≥99% beta-tubulin sequence similarity and supported by 97% bootstrap value) and four isolates to the genus level (Table S3 and Figure 3). All the isolates belonged to the phylum Ascomycota within the Aspergillaceae, Cladosporiaceae, and Pleosporaceae families (Figure 3). *Alternaria* sp. was isolated from the Perdido region (station N2; ID:4), whereas *C. halotolerans* (station C11; ID:1), *Penicillium* sp. 1 (station D16; ID:5), *Penicillium* sp. 2 (station C11; ID:6), and *Stemphylium* sp. (station C12; ID:3) were isolated from the Coatzacoalcos region (Table S3).

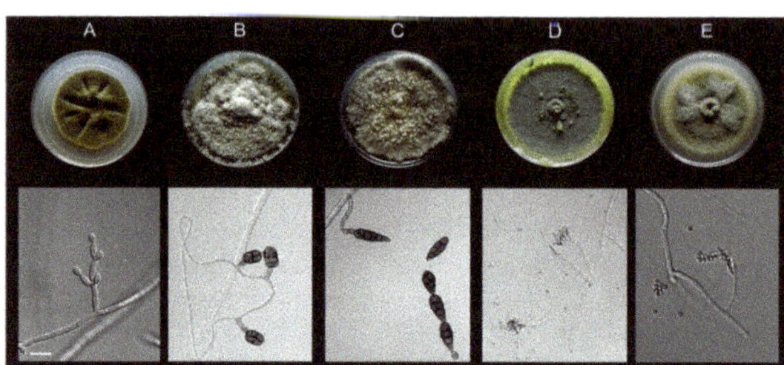

Figure 2. Macromorphology and micromorphology of the fungal isolates recovered from deep-sea sediment samples evaluated for their capacity to degrade petroleum. Top row: colony morphology of strains grown on Petri dishes containing potato dextrose agar medium. Bottom row: conidiophores and conidia of the fungal strains imaged by Differential Interference Contrast microscopy. Scale bar = 10 µm. (**A**) *Cladosporium halotolerans*, (**B**) *Stemphylium* sp., (**C**) *Alternaria* sp., (**D**) *Penicillium* sp. 1, and (**E**) *Penicillium* sp. 2.

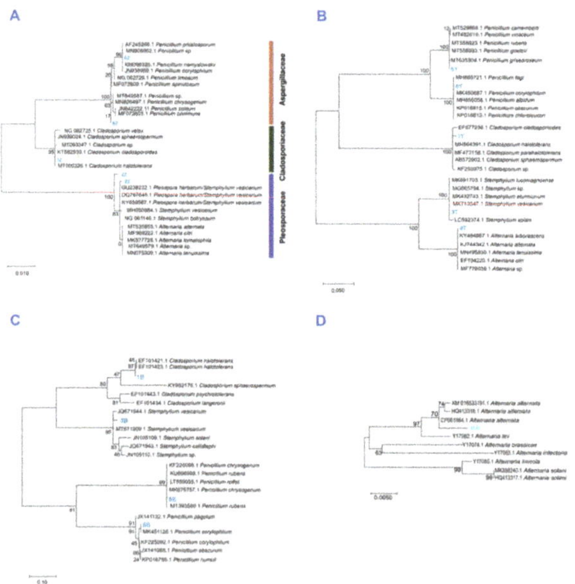

Figure 3. Evolutionary analysis by Maximum Likelihood method for (**A**) 18S, (**B**) ITS, and (**C**) β-tubulin. Bootstrap values are indicated above the branches. The phylogenetic position of deep-sea derived fungal sequences is denoted by blue color. The evolutionary history was inferred by using K2+I (**A**,**B**), K2+G (**C**), and HKY+G (**D**) substitution models. The tree with the highest log likelihood is shown. The initial tree for the heuristic search was obtained automatically by applying the Neighbor-Join and BioNJ algorithms to a matrix of pairwise distances estimated using the Maximum Composite Likelihood (MCL) approach and then selecting the topology with the superior log likelihood value. The tree is drawn to scale, with the branch lengths measured in the number of substitutions per site.

3.2. *Alternaria sp. Was Able to Grow with Extra-Heavy Crude Oil as the Sole Carbon Source*

The five fungal strains were assayed for growth at 0.5% of three different crude oils: light, heavy, or extra-heavy. All five strains were able to grow in light crude oil as the sole source of carbon. However, the heavier the crude oil was, the fewer strains were able to metabolize it. While four strains (*Alternaria* sp., *Penicillium* sp. 1, *Penicillium* sp. 2, and *Stemphylium* sp.) were able to grow in HCO, only one, *Alternaria* sp., was able to grow in EHCO (Table 1).

Table 1. Growth of selected fungi in crude oils of different densities. The culture was carried out in 50 mL of modified Czapek minimal medium with 0.5% *w/v* of crude oil as the sole carbon source.

Identified Isolate	Crude Oil Density		
	Light 40° API	**Heavy 16–20° API**	**Extra-Heavy 7–10° API**
Alternaria sp.	+	+	+
Cladosporium halotolerans	+	−	−
Penicillium sp. 1	+	+	−
Penicillium sp. 2	+	+	−
Stemphylium sp.	+	+	−

+ growth, − no growth.

Under light microscopy, *Alternaria* sp., *Penicillium* sp. 1, and *Penicillium* sp. 2 hyphae were observed strongly adhered to the surface of the HCO (Figure 4A–C). In contrast, *Stemphylium* sp. presented hyphal growth that did not adhere to the oil (Figure 4D). Regardless of how attached the hyphae were to the crude oil, the main purpose of this experiment was to confirm the presence of mycelium, and this was successfully achieved, as shown in the images obtained. When grown in EHCO, hyphae and conidia of *Alternaria* sp. also adhered to the sticky oil drops. By DIC microscopy, what appeared to be newly formed conidia at the tips of the hyphae were observed at the petroleum surface (Figure 5). By SEM, a dense mass of mycelium and conidia were observed as well growing at the surface of the petroleum (Figure 6).

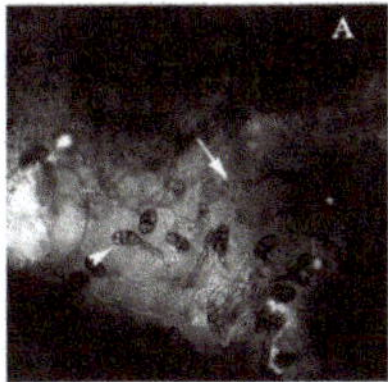

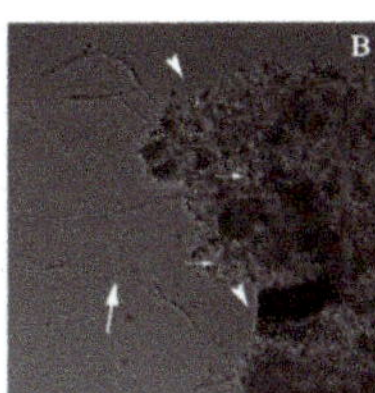

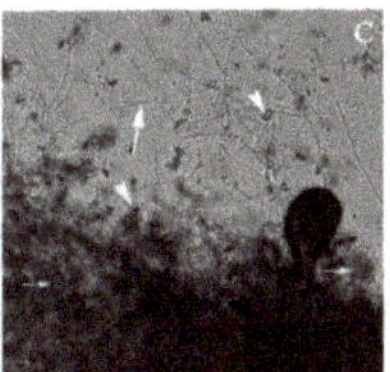

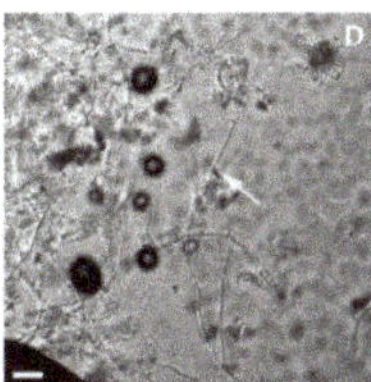

Figure 4. Differential interference contrast microscopy of fungal mycelia grown in Czapek minimal medium containing 0.5% *w/v* of heavy crude oil (16–20° API). (**A**) *Alternaria* sp. conidia (arrowheads) and mycelium (arrows) adhered to the oil after six days of incubation. (**B**) *Penicillium* sp. 1 hyphae (arrows) and conidia growth (arrowheads) after 24 days of incubation; salts from the Czapek minimal medium (small arrows) can be observed. (**C**) *Penicillium* sp. 2 mycelium (arrows) and conidia (arrowheads) embedded in salt (small arrows) and petroleum after 24 days of incubation. (**D**) *Stemphylium* sp. hyphae (arrows) with no attachment to petroleum after 27 days of incubation. Scale bar = 25 µm.

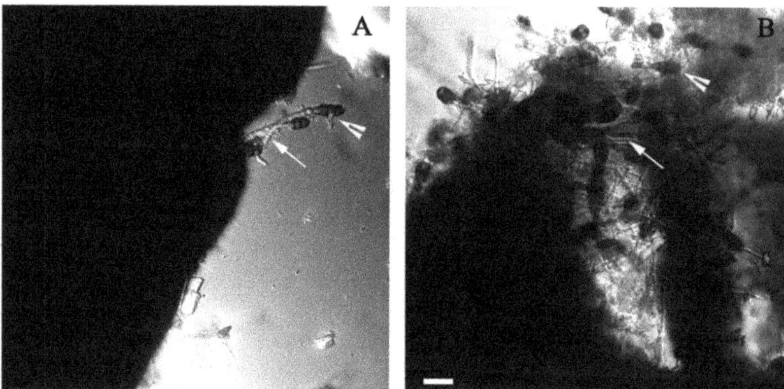

Figure 5. Differential interference contrast microscopy of *Alternaria* sp. grown in Czapek minimal medium containing 0.5% *w/v* of extra-heavy crude oil (7–10° API). (**A**) Hyphae (arrow) and septate conidia (arrowhead) emerging from a dense mass of petroleum. (**B**) Mycelium (arrow) and conidia (arrowhead) with Czapek minimal medium salts and petroleum. Scale bar = 25 μm.

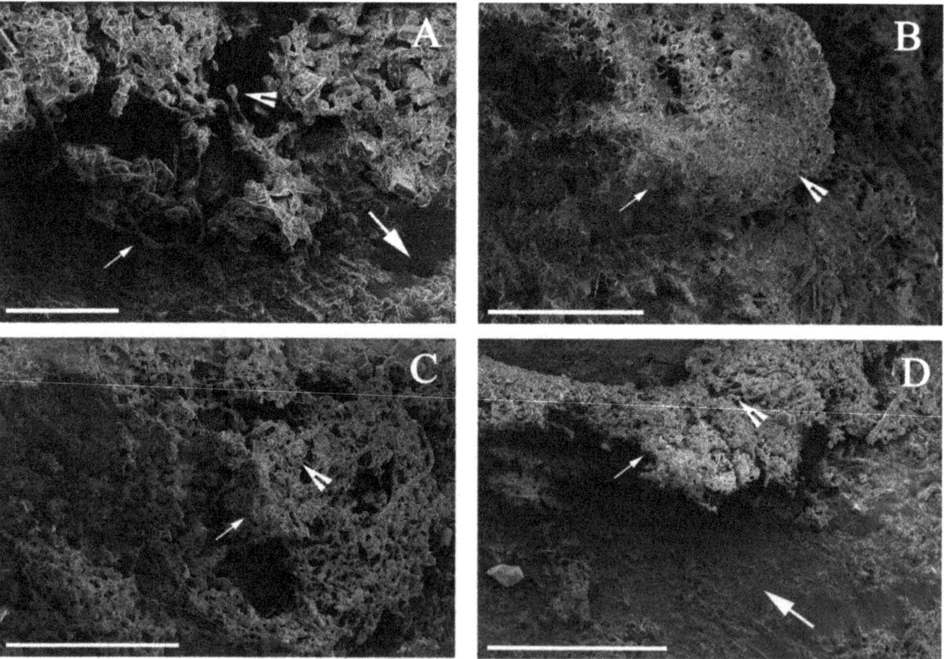

Figure 6. Scanning electron microscopy of *Alternaria* sp. grown in Czapek minimal medium containing 0.5% *w/v* of extra-heavy crude oil (7–10° API). (**A**) Hyphae (small arrows), conidia (arrowheads), and smooth petroleum surface (arrows). Scale bar = 100 μm. (**B**) Mycelium (small arrows) and conidia (arrowheads). Scale bar = 200 μm. (**C**) Hyphae (small arrows) and conidia (arrowheads). Scale bar = 300 μm. (**D**) Mycelium (small arrows), conidia (arrowheads), and petroleum (arrows). Scale bar = 500 μm. In all images, the mycelium is in tight contact with the petroleum surface.

For both *Penicillium* spp., the maximum average amount of protein (40.5 ± 1.2 mg/L for *Penicillium* 1 and 46.0 ± 13.7 mg/L for *Penicillium* 2) was obtained for the cultures amended with 0.75% *v/v* light crude oil (Figure S1). The protein concentration in *Penicillium* sp. 1 was significantly ($p = 0.012$) higher at the 0.75% (*v/v*) concentration than 0.25% (*v/v*) of the light crude oil. Whereas, for *Penicillium* sp. 2, no significant differences could be observed at the different concentrations of light crude oil ($p = 5.14$) (Figure S1).

To quantify protein concentration in the cultures of *Alternaria* sp. grown in Czapek minimal medium supplemented with 0.5% *v/v* EHCO, it was necessary to add toluene to separate the mycelium from the oil, because there was a strong attachment between them, and the mycelium could not be easily recovered through filtration. An average total protein increase of 7.04 ± 0.20 mg/L was obtained for *Alternaria* spp. cultures that were grown for a month in Czapek minimal medium supplemented with 0.5% *w/v* EHC. That value was slightly higher, although not more significantly than the one obtained for the control cultures without EHCO (5.9 ± 0.32 mg/L) ($p = 0.066$).

3.3. Alternaria sp. Mainly Metabolizes the Aromatic Components of the EHCO

After 30 days of growth, EHCO degradation by *Alternaria* sp. was determined by gas chromatography in both whole EHCO and their fractions (Figure 7). The main components of the original EHCO corresponded to the aromatic fraction with 43.5% (±10.8), followed by saturates with 35.5% (±1.5) and the polar (resins-asphaltenes) fractions 21% (±1.9).

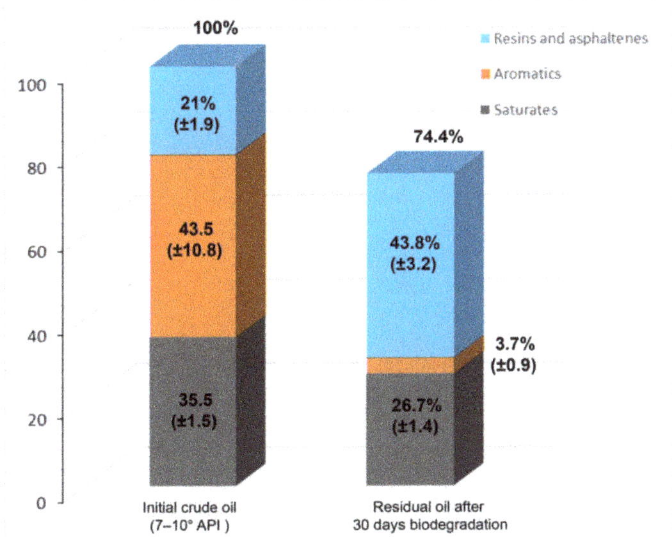

Figure 7. Hydrocarbon degradation after 30 days of growth by *Alternaria* sp. with extra-heavy crude oil (7–10° API) as the sole source of carbon initial and residual oil compositions. The number on the top of the columns represents the residual oil after biodegradation as a percentage of initial crude oil. The numbers on each column are the percentage of each fraction. The oils were fractionated by the modified SARA procedure to obtain the saturates, aromatics, and polar (resins and asphaltenes) fractions. The total petroleum hydrocarbons were determined by gas chromatography equipped with a flame ionization detector.

According to the chromatograms, *Alternaria* sp. degraded 25.6% (±7.6) of the total crude oil (Figure S2). Noteworthy, *Alternaria* sp. had a higher metabolization of the aromatic fraction, with 91.3% (±15.7) of degradation of these compounds. It also degraded the saturates fraction by 24.7% (±1.0). The resin–asphaltene fractions showed an increase of 108% (±5.8), mainly due to the production of polar molecules. In other words, the composition of the crude oil after biodegradation was 26.7% (±1.4) of the saturates fraction, 3.7% (±0.9) of the aromatic fraction, and 43.8% (±3.2) of polar (resin–asphaltene) fractions (Figure 7).

4. Discussion

Following the 2010 Deepwater Horizon oil spill, fungal strains from oil-soaked sand patties collected from beaches [41] and bacteria from oiled beach sands at Pensacola Beach, FL, USA [42], both sites located in the American EEZ, were identified as potential bioremediation microorganisms. Fungi synthesize a battery of enzymes, including laccases, peroxidases, lytic polysaccharide monooxygenases, and monooxygenase cytochrome P450 that degrade different and complex substrates, including PAHs [43,44]. In addition, isolates from extreme environments have been proven to produce extracellular extremophilic ligninolytic enzymes, which can perform catalytic activities under adverse environments without being denatured [45]. These metabolic capacities and rapid mycelial propagation pose fungi as attractive prospects for bioremediation applications [46]. Notably, marine-derived fungi have been recognized as promising bioremediation agents due to their metabolic adaptations to naturally occurring oil sources [47,48]. Yet, they have been scarcely studied for their ability to use recalcitrant elements of HCO and EHCO.

In this study, we showed the ability of *Alternaria* sp., *Penicillium* spp., and *Stemphylium* sp., isolates recovered from deep-sea sediments of the Mexican EEZ of the Gulf of Mexico to grow using HCO and, most importantly, the ability of *Alternaria* sp. to grow using EHCO as sole carbon source. Unfortunately, given the nature of the EHCO and the strong attachment of *Alternaria* sp. mycelium to the EHCO in the glass rod, the quantification of the biomass was not successfully attained. However, at the visible level, one could observe the mycelium of *Alternaria* sp. only in the cultures supplemented with EHCO (Figure S3). We concluded that the toluene added for protein quantification to get an indirect measurement of the fungal biomass probably interfered with the complete recovery of the mycelium, and the protein content obtained was not an accurate estimate of the actual amount of protein. Although the fungal taxa identified in this study have been previously reported as light and medium crude oil degraders, there are no previous reports of HCO or EHCO degradation by any of them. *Penicillium* spp. isolated from diverse polluted environments have been extensively reported as capable of removing a wide array of contaminants, including heavy metals and petroleum hydrocarbons [49,50]. *Penicillium* spp. can efficiently degrade polycyclic aromatic hydrocarbons such as fluorene [51], pyrene [52], and phenanthrene [53]. The *Stemphylium* sp. has been isolated from petroleum-contaminated soils in Iran [54] and from a Mediterranean marine site after an oil spill, where *C. halotolerans* was recovered as well [55]. *P. corylophilum* was isolated from ship-breaking yards of Bangladesh and was able to degrade petroleum hydrocarbon [56].

Alternaria spp. have been isolated from terrestrial contaminated zones with evidence of PAH degradation [57], from tar balls with a very limited ability to degrade crude oil [49], and from noncontaminated gardens showing the ability to degrade octane and decane [58]. *A. alternata* has the capacity to synthesize enzymes related to hydrocarbon degradation, such as manganese peroxidase and lignin peroxidase [59,60]. There is a comparative analysis of the secretome profile of Manganese (II)-Oxidizing [61], as well as a report of lignin peroxidase production [62], and this sets a precedent for the potential degradation capacity of *Alternaria* sp.

Heavy crude oils contain a high proportion of high molecular weight hydrocarbons and elevated levels of hetero compounds, including sulfur, nitrogen, oxygen, and heavy metals [63]. The typical SARA composition of heavy oils is 4–23% saturates, 14–32% aromatics, 27–46% resins, and 15–43% asphaltenes [64]. The high concentration of resins and asphaltenes, together with the high concentration of high molecular weight PAHs in EHCO, poses a challenge for biodegradation and bioremediation processes. The chemical nature of the products from hydrocarbon degradation was not analyzed in this study. The composition of crude oil includes several thousands of different hydrocarbons that could potentially represent a carbon source for *Alternaria* sp. The products resulting from microbial degradation by *Alternaria* are also a complex mixture of polar compounds hard to resolve by conventional analytical chemistry tools. However, hydrocarbon degradation products' chemical natures have been extensively studied [65,66]. Different enzymes transform saturated and aromatic hydrocarbons, increasing the relative abundance of polar fractions. Various pathways, such as terminal oxidation, subterminal oxidation, ω-oxidation, and β-oxidation, are involved in the degradation of aliphatic hydrocarbons. The initial alkane oxidation results in the formation of an alcohol, which can then be metabolized by the β-oxidation pathway of fatty acids. On the other hand, fungi's initial oxidation of aromatic hydrocarbons is mediated by monooxygenases forming a trans-diol. The benzene ring is then cleaved in different ways by appropriate enzymes leading to the formation of central intermediates, which are further converted to tricarboxylic acid (TCA) cycle intermediates. Finally, cyclic alkanes are converted to cyclic alcohols and dehydrogenated to ketones by an oxidase system. Then, a monooxygenase system forms lactonates, and a lactone hydrolase finally opens the ring.

Our study proved that *Alternaria* sp. has a great affinity for the aromatic fraction of EHCO, showing an extensive transformation of 91.3% of this fraction. The aromatic fraction that was transformed by *Alternaria* sp. to more polar compounds migrated to the resin–asphaltene fraction in the degraded hydrocarbons, as suggested by the percentage increase of this fraction. Despite this increase, it has been previously well-established that the polar compounds appearing in the resin–asphaltene fraction as a result of the biodegradation of saturates and, especially, of the aromatic fraction, are more biodegradable than the original hydrocarbons, and they can be more accessible for degradation by other microorganisms [67,68].

The station N2 (Perdido region), from where *Alternaria* sp. was isolated, did not display any remarkable difference in salinity, temperature, or dissolved oxygen concentration compared with the rest of the stations (Supplementary Table S2). The Gulf of Mexico has a two-layer system; the upper layer (above 800–1000 m) is highly regulated by the Loop Current and its associated eddies (LCEs) entering the Gulf of Mexico through the Yucatan Channel [69]. The river discharges, the temperature variation, and the freshwater input do not impact the salinity or temperature beyond the upper 100 m of the water column [70]. Consequently, the main driver in the hydrographic conditions in the GoM where the sediments samples were taken was the local presence of LCEs [70]. Regardless of the wide distribution of the sites sampled in the Gulf of Mexico, it is not possible to recognize a difference between the Perdido and Coatzacoalcos Stations. Temperature and dissolved oxygen variation respond mainly to the depth of the sediments, and there is an insignificant variation in the salinity in all the stations.

Recent estimations report that unconventional oil reserves, which include HCO, EHCO, and bitumen, could exceed six trillion barrels [71]. The ineludible extraction of these reservoirs will inevitably produce oil spills. It is essential to explore native fungal microorganisms capable of degrading HCO and, especially, EHCO to mitigate the environmental impact of accidental petroleum spills. Offshore oil spills are of tremendous concern due to their potential impact on the economy and especially on ecological systems [72]. The successful bioremediation of marine oil spills has been reported [73]. Among the bioremediation strategies, biostimulation is an *in situ* technology that involves the introduction of indigenous (obtained from the contaminated site) or exogenous hydrocarbonoclastic microorganisms to a polluted site to enhance microbial degradation at the site. Many degradation studies are focused on water column degradation, and no attention has been given to sediment oil degradation. Oil biodegradation in sediments is significant for the heavy fractions of oil, which tend to settle after the oil spill. In addition, the sediment environmental conditions are different such as oxygen and nutrient concentrations. There are nevertheless some challenges associated with bioremediation of spilled petroleum [74], including: (i) the resistance of heavy fractions, especially asphaltenes and high molar mass polycyclic aromatic hydrocarbons, to degradation. Thus, it seems crucial to have microbial strains able to perform the biodegradation of heavy and extra-heavy oils such as *Alternaria* sp. isolated and characterized in this work; (ii) eutrophication caused by biostimulation. Seawater is characterized by the low nitrogen and phosphorous concentrations that are limiting factors for oil degradation. The use of oleophilic fertilizers seems to be an effective nutrient type to be applied in marine environments to prevent eutrophication; (iii) unsustainability of bioaugmentation in the field. This can be reduced using well-adapted microorganisms, like, for instance, those adapted to sediment conditions. In addition, the encapsulation of the microorganism can protect it in the new environmental conditions; (iv) poor bioavailability of spilled petroleum. Oil biodegradation takes place at the oil–water interphase. Therefore, the widely used oil dispersion or solubilization by adding chemical or biological surfactants enhances biodegradation. However, the chemical agents could be potentially toxic to the environment; and (v) oil biodegradation inefficiency in semi-anoxic environments. As mentioned before, in deep sediments, there is a low oxygen concentration. It is also important to have microorganisms isolated from similar sediment conditions, such as the *Alternaria* sp. isolated and characterized in this work.

Considering all the above, microorganisms that, like *Alternaria* sp., are capable of degrading some of the heavier and more recalcitrant components of these crude oils can be good candidates for further studies testing different environmental conditions in order to assess their use for bioremediation in the event of an oil spill.

Supplementary Materials: The following are available online at https://www.mdpi.com/article/10.3390/app11136090/s1: Table S1: Geolocalization and data of stations sampled in the Gulf of Mexico. Table S2: Sampling stations and locations from where the fungal strains were isolated. Table S3: Reference DNA sequences of the ITS rDNA, 18S and ß-tubulin markers used in the phylogenetic analyses. Figure S1: Protein content (average of three replicates) of *Penicillium* sp. 1 (A) and *Penicillium* sp. 2 (B) grown under different light crude oil concentrations. Figure S2: Chromatograms of whole extra-heavy crude oil and its saturates, aromatics and polars (SAP) fractions. Figure S3: *Alternaria* sp. grown with (B, D) or without (A, C) 0.5% w/v of extra-heavy crude oil as sole carbon source after a month of incubation.

Author Contributions: Conceptualization, M.R.; methodology, L.R.-H., M.D.C.-L., P.V. and I.B.-G.; investigation, L.R.-H.; writing—original draft preparation, L.R.-H.; writing—review and editing, M.R., P.V. and R.V.-D.; formal analysis, L.R.-H.; supervision, M.R. and R.V.-D.; project administration, M.R.; and funding acquisition, M.R. All authors have read and agreed to the published version of the manuscript.

Funding: This research was funded by the National Council of Science and Technology of Mexico-Mexican Ministry of Energy-Hydrocarbon Trust, project 201441. This is a contribution of the Gulf of Mexico Research Consortium (CIGoM). LRH was supported by CONACYT doctoral fellowship 259695.

Institutional Review Board Statement: Not applicable.

Informed Consent Statement: Not applicable.

Data Availability Statement: The data is contained within the article and supplementary material.

Acknowledgments: We thank Karla Sidon for collecting the samples and Lluvia Vargas-Gastélum for carefully reading the manuscript. Microscopy analyses were carried out at the National Laboratory of Advanced Microscopy of CICESE (LNMA) and the Department of Optics of CICESE. We are grateful to Diego Delgado, Fabian Cordero, Pilar Sánchez-Saavedra, and Juan Manuel Martínez for their technical support.

Conflicts of Interest: The authors declare that they have no known competing financial interests or personal relationships that could have appeared to influence the work reported in this paper.

Abbreviations

Gulf of Mexico	GoM
Heavy Crude Oil	HCO
Extra-heavy Crude Oil	EHCO
Polycyclic Aromatic Hydrocarbons	PAHs
Metagenomics Campaign	MET-II
Potato Dextrose Agar	PDA
Differential Interference Contrast	DIC
Internal Transcribed Spacer	ITS
β-tubulin gene	benA
Maximum Likelihood	ML
Maximum Composite Likelihood	MCL
American Petroleum Institute	API
Total Petroleum Hydrocarbons	TPH
Flame Ionization Detector	FID
Scanning Electron Microscopy	SEM
Saturate, Aromatic and Polar	SAP
Economic Exclusive Zone	EZZ
Gulf of Mexico Research Consortium	CIGoM

References

1. Matsuo, Y.; Yanagisawa, A.; Yamashita, Y. A global energy outlook to 2035 with strategic considerations for Asia and Middle East energy supply and demand interdependencies. *Energy Strat. Rev.* **2013**, *2*, 79–91. [CrossRef]
2. British Petroleum Company. *BP Energy Outlook 2035*; British Petroleum Company: London, UK, 2015.
3. Kuppusamy, S.; Maddela, N.R.; Megharaj, M.; Venkateswarlu, K. Total Petroleum Hydrocarbons. *Environ. Fate Toxic. Remediat.* **2020**. [CrossRef]
4. Guo, K.; Li, H.; Yu, Z. In situ heavy and extra-heavy oil recovery: A review. *Fuel* **2016**, *185*, 886–902. [CrossRef]
5. Demirbas, A.; Bafail, A.; Nizami, A.-S. Heavy oil upgrading: Unlocking the future fuel supply. *Pet. Sci. Technol.* **2016**, *34*, 303–308. [CrossRef]
6. Covert, T.; Greenstone, M.; Knittel, C.R. Will We Ever Stop Using Fossil Fuels? *J. Econ. Perspect.* **2016**, *30*, 117–138. [CrossRef]
7. He, L.; Lin, F.; Li, X.; Sui, H.; Xu, Z. Interfacial sciences in unconventional petroleum production: From fundamentals to ap-plications. *Chem. Soc. Rev.* **2015**, *44*, 5446–5494. [CrossRef]
8. Farrington, J.W. Need to update human health risk assessment protocols for polycyclic aromatic hydrocarbons in seafood after oil spills. *Mar. Pollut. Bull.* **2020**, *150*, 110744. [CrossRef] [PubMed]
9. Warnock, A.M.; Hagen, S.C.; Passeri, D.L. Marine Tar Residues: A Review. *Water Air Soil Pollut.* **2015**, *226*, 1–24. [CrossRef]
10. Zhang, B.; Matchinski, E.J.; Chen, B.; Ye, X.; Jing, L.; Lee, K. Marine Oil Spills—Oil Pollution, Sources and Effects. In *World Seas: An Environmental Evaluation*; Elsevier BV: Amsterdam, The Netherlands, 2019; pp. 391–406.
11. Harwell, M.A.; Gentile, J.H. Ecological significance of residual exposures and effects from the Exxon Valdez oil spill. *Integr. Environ. Assess. Manag.* **2006**, *2*, 204. [CrossRef]
12. Mei, H.; Yin, Y. Studies on marine oil spills and their ecological damage. *J. Ocean. Univ. China* **2009**, *8*, 312–316. [CrossRef]
13. Das, N.; Chandran, P. Microbial Degradation of Petroleum Hydrocarbon Contaminants: An Overview. *Biotechnol. Res. Int.* **2011**, *2011*, 1–13. [CrossRef]
14. Frias-Lopez, J.; Shi, Y.; Tyson, G.; Coleman, M.; Schuster, S.C.; Chisholm, S.W.; DeLong, E.F. Microbial community gene expression in ocean surface waters. *Proc. Natl. Acad. Sci. USA* **2008**, *105*, 3805–3810. [CrossRef] [PubMed]
15. Al-Hawash, A.B.; Dragh, M.A.; Li, S.; Alhujaily, A.; Abbood, H.A.; Zhang, X.; Ma, F. Principles of microbial degradation of petroleum hydrocarbons in the environment. *Egypt. J. Aquat. Res.* **2018**, *44*, 71–76. [CrossRef]
16. Naranjo-Briceño, L.; Pernía, B.; Perdomo, T.; González, M.; Inojosa, Y.; De Sisto, Á.; Urbina, H.; León, V. Potential role of extremophilic hydrocarbonoclastic fungi for extra-heavy crude oil bioconversion and the sustainable development of the petroleum industry. In *Fungi in Extreme Environments: Ecological Role and Biotechnological Significance*; Tiquia-Arashiro, S.M., Grube, M., Eds.; Springer: Cham, Switzerland, 2019; pp. 559–586. [CrossRef]
17. Uribe-Alvarez, C.; Ayala, M.; Perezgasga, L.; Naranjo, L.; Urbina, H.; Vazquez-Duhalt, R. First evidence of mineralization of petroleum asphaltenes by a strain of *Neosartorya fischeri*. *Microb. Biotechnol.* **2011**, *4*, 663–672. [CrossRef]
18. López, E.L.H.; Perezgasga, L.; Huerta-Saquero, A.; Mouriño-Pérez, R.R.; Vazquez-Duhalt, R. Biotransformation of petroleum asphaltenes and high molecular weight polycyclic aromatic hydrocarbons by *Neosartorya fischeri*. *Environ. Sci. Pollut. Res.* **2016**, *23*, 10773–10784. [CrossRef]
19. Naranjo-Briceño, L.; Pernia, B.; Guerra, M.; Demey, J.R.; De Sisto, Á.; Inojosa, Y.; Gonzalez, M.; Fusella, E.; Freites, M.; Yegres, F. Potential role of oxidative exoenzymes of the extremophilic fungus *Pestalotiopsis palmarum* BM-04 in biotransformation of extra-heavy crude oil. *Microb. Biotechnol.* **2013**, *6*, 720–730. [CrossRef]

20. Chaillan, F.; Le Flèche, A.; Bury, E.; Phantavong, Y.-H.; Grimont, P.; Saliot, A.; Oudot, J. Identification and biodegradation potential of tropical aerobic hydrocarbon-degrading microorganisms. *Res. Microbiol.* **2004**, *155*, 587–595. [CrossRef]
21. Ayala, M.; López, E.L.H.; Perezgasga, L.; Vazquez-Duhalt, R. Reduced coke formation and aromaticity due to chloroperoxidase-catalyzed transformation of asphaltenes from Maya crude oil. *Fuel* **2012**, *92*, 245–249. [CrossRef]
22. Pourfakhraei, E.; Badraghi, J.; Mamashli, F.; Nazari, M.; Saboury, A.A. Biodegradation of asphaltene and petroleum compounds by a highly potent Daedaleopsi ssp. *J. Basic Microbiol.* **2018**, *58*, 609–622. [CrossRef] [PubMed]
23. Murawski, S.A.; Hollander, D.J.; Gilbert, S.; Gracia, A. Deepwater Oil and Gas Production in the Gulf of Mexico and Related Global Trends. In *Scenarios and Responses to Future Deep Oil Spills*; Springer: Berlin/Heidelberg, Germany, 2020; pp. 16–32.
24. Velez, P.; Gasca-Pineda, J.; Riquelme, M. Cultivable fungi from deep-sea oil reserves in the Gulf of Mexico: Genetic signatures in response to hydrocarbons. *Mar. Environ. Res.* **2020**, *153*, 104816. [CrossRef] [PubMed]
25. Hickey, P.C.; Swift, S.R.; Roca, M.G.; Read, N.D. Live-cell imaging of filamentous fungi using vital fluorescent dyes and con-focal microscopy. *Methods Microbiol.* **2004**, *34*, 63–87. [CrossRef]
26. Kohlmeyer, J.; Kohlmeyer, E. *Marine Mycology: The Higher Fungi*; Academic Press: Cambridge, MA, USA, 1979.
27. Lücking, R.; Aime, M.C.; Robbertse, B.; Miller, A.N.; Ariyawansa, H.A.; Aoki, T.; Cardinali, G.; Crous, P.W.; Druzhinina, I.S.; Geiser, D.M.; et al. Unambiguous identification of fungi: Where do we stand and how accurate and precise is fungal DNA barcoding? *IMA Fungus* **2020**, *11*, 1–32. [CrossRef] [PubMed]
28. Gardes, M.; Bruns, T.D. ITS primers with enhanced specificity for basidiomycetes—Application to the identification of mycorrhizae and rusts. *Mol. Ecol.* **1993**, *2*, 113–118. [CrossRef]
29. White, T.; Bruns, T.; Lee, S.; Taylor, J.; Innis, M.; Gelfand, D.; Sninsky, J. Amplification and direct sequencing of fungal ribo-somal RNA genes for phylogenetics. In *PCR Protocols: A Guide to Methods and Applications*; Academic Press: Cambridge, MA, USA, 1990.
30. Gargas, A.; Taylor, J.W. Polymerase chain reaction (PCR) primers for amplifying and sequencing nuclear 18S rDNA from lichenized fungi. *Mycologia* **1992**, *84*, 589–592. [CrossRef]
31. Glass, N.L.; Donaldson, G.C. Development of primer sets designed for use with the PCR to amplify conserved genes from filamentous ascomycetes. *Appl. Environ. Microbiol.* **1995**, *61*, 1323–1330. [CrossRef]
32. Ewing, B.; Hillier, L.; Wendl, M.C.; Green, P. Base-Calling of Automated Sequencer Traces UsingPhred. I. Accuracy Assessment. *Genome Res.* **1998**, *8*, 175–185. [CrossRef]
33. Ewing, B.; Green, P. Base-calling of automated sequencer traces using Phred. II. Error probabilities. *Genome Res.* **1998**, *8*, 186–194. [CrossRef] [PubMed]
34. Gordon, D.; Desmarais, C.; Green, P. Automated Finishing with Autofinish. *Genome Res.* **2001**, *11*, 614–625. [CrossRef] [PubMed]
35. Okonechnikov, K.; Golosova, O.; Fursov, M. The UGENE Team. Unipro UGENE: A unified bioinformatics toolkit. *Bioinformatics* **2012**, *28*, 1166–1167. [CrossRef]
36. Edgar, R.C. MUSCLE: Multiple sequence alignment with high accuracy and high throughput. *Nucleic Acids Res.* **2004**, *32*, 1792–1797. [CrossRef] [PubMed]
37. Stecher, G.; Tamura, K.; Kumar, S. Molecular Evolutionary Genetics Analysis (MEGA) for macOS. *Mol. Biol. Evol.* **2020**, *37*, 1237–1239. [CrossRef] [PubMed]
38. Kumar, S.; Stecher, G.; Li, M.; Knyaz, C.; Tamura, K. MEGA X: Molecular evolutionary genetics analysis across computing platforms. *Mol. Biol. Evol.* **2018**, *35*, 1547–1549. [CrossRef]
39. Kimura, M. A simple method for estimating evolutionary rates of base substitutions through comparative studies of nucleotide sequences. *J. Mol. Evol.* **1980**, *16*, 111–120. [CrossRef]
40. Abalos, A.; Viñas, M.; Sabate, J.; Manresa, A.; Solanas, A. Enhanced Biodegradation of Casablanca Crude Oil by A Microbial Consortium in Presence of a Rhamnolipid Produced by Pseudomonas Aeruginosa AT10. *Biodegradation* **2004**, *15*, 249–260. [CrossRef] [PubMed]
41. Simister, R.; Poutasse, C.; Thurston, A.; Reeve, J.; Baker, M.; White, H. Degradation of oil by fungi isolated from Gulf of Mexico beaches. *Mar. Pollut. Bull.* **2015**, *100*, 327–333. [CrossRef] [PubMed]
42. Kostka, J.E.; Prakash, O.; Overholt, W.A.; Green, S.; Freyer, G.; Canion, A.; Delgardio, J.; Norton, N.; Hazen, T.C.; Huettel, M. Hydrocarbon-Degrading Bacteria and the Bacterial Community Response in Gulf of Mexico Beach Sands Impacted by the Deepwater Horizon Oil Spill. *Appl. Environ. Microbiol.* **2011**, *77*, 7962–7974. [CrossRef]
43. Otero-Blanca, A.; Folch-Mallol, J.L.; Lira-Ruan, V.; del Rayo Sánchez-Carbente, M.; Batista-García, R.A. Phytoremediation and Fungi: An Underexplored Binomial. In *Approaches in Bioremediation*; Springer Science and Business Media LLC: Berlin/Heidelberg, Germany, 2018; pp. 79–95.
44. Deshmukh, R.; Khardenavis, A.A.; Purohit, H.J. Diverse Metabolic Capacities of Fungi for Bioremediation. *Indian J. Microbiol.* **2016**, *56*, 247–264. [CrossRef]
45. Chandra, R.; Kumar, V.; Yadav, S. Extremophilic Ligninolytic Enzymes. In *Extremophilic Enzymatic Processing of Lignocellulosic Feedstocks to Bioenergy*; Springer Science and Business Media LLC: Berlin/Heidelberg, Germany, 2017; pp. 115–154.
46. Vidali, M. Bioremediation. An overview. *Pure Appl. Chem.* **2001**, *73*, 1163–1172. [CrossRef]
47. Kumar, R.; Kaur, A. Oil spill removal by mycoremediation. In *Microbial Action on Hydrocarbons*; Kumar, V., Kumar, M., Pra-sad, R., Eds.; Springer: Singapore, 2018; pp. 505–526. [CrossRef]
48. Kvenvolden, K.A.; Cooper, C.K. Natural seepage of crude oil into the marine environment. *Geo-Marine Lett.* **2003**, *23*, 140–146. [CrossRef]

49. Elshafie, A.; Alkindi, A.Y.; Al-Busaidi, S.; Bakheit, C.; Albahry, S. Biodegradation of crude oil and n-alkanes by fungi isolated from Oman. *Mar. Pollut. Bull.* **2007**, *54*, 1692–1696. [CrossRef]
50. Singh, H. *Mycoremediation: Fungal Bioremediation*; John Wiley & Sons: Hoboken, NJ, USA, 2006.
51. Garon, D.; Sage, L.; Seigle-Murandi, F. Effects of fungal bioaugmentation and cyclodextrin amendment on fluorene degrada-tion in soil slurry. *Biodegradation* **2004**, *15*, 1–8. [CrossRef] [PubMed]
52. Saraswathy, A.; Hallberg, R. Mycelial pellet formation by *Penicillium ochrochloron* species due to exposure to pyrene. *Microbiol. Res.* **2005**, *160*, 375–383. [CrossRef]
53. Melendez-Estrada, J.; Amezcua-Allieri, M.A.; Alvarez, P.; Rodríguez-Vázquez, R. Phenanthrene Removal byPenicillium frequen-tansGrown on a Solid-State Culture: Effect of Oxygen Concentration. *Environ. Technol.* **2006**, *27*, 1073–1080. [CrossRef]
54. Mohammadian, E.; Arzanlou, M.; Babai-Ahari, A. Diversity of culturable fungi inhabiting petroleum-contaminated soils in Southern Iran. *Antonie van Leeuwenhoek* **2017**, *110*, 903–923. [CrossRef] [PubMed]
55. Bovio, E.; Gnavi, G.; Prigione, V.; Spina, F.; Denaro, R.; Yakimov, M.; Calogero, R.; Crisafi, F.; Varese, G.C. The culturable mycobiota of a Mediterranean marine site after an oil spill: Isolation, identification and potential application in bioremediation. *Sci. Total Environ.* **2017**, *576*, 310–318. [CrossRef] [PubMed]
56. Dhar, K.; Dutta, S.; Anwar, M.N. Biodegradation of petroleum hydrocarbon by indigenous fungi isolated from ship breaking yards of Bangladesh. *Int. Res. J. Biol. Sci.* **2014**, *3*, 22–30.
57. Balaji, V.; Arulazhagan, P.; Ebenezer, P. Enzymatic bioremediation of polyaromatic hydrocarbons by fungal consortia en-riched from petroleum contaminated soil and oil seeds. *J. Environ. Biol.* **2014**, *35*, 521–529. [PubMed]
58. Loretta, O.O.; Samuel, O.; Johnson, G.H.; Emmanuel, S. Comparative Studies on the Biodegradation of Crude Oil-polluted Soil by Pseudomonas aeruginosa and Alternaria Species Isolated from Unpolluted Soil. *Microbiol. Res. J. Int.* **2017**, *19*, 1–10. [CrossRef]
59. Ali, M.I.; Khalil, N.M.; Abd El-Ghany, M.N. Biodegradation of some polycyclic aromatic hydrocarbons by *Aspergillus terreus*. *Afr. J. Microbiol. Res.* **2012**, *6*, 3783–3790.
60. Vazquez-Duhalt, R.; Westlake, D.W.S.; Fedorak, P.M. Lignin peroxidase oxidation of aromatic compounds in systems con-taining organic solvents. *Appl. Environ. Microbiol.* **1994**, *60*, 459–466. [CrossRef] [PubMed]
61. Zeiner, C.A.; Purvine, S.; Zink, E.M.; Paša-Tolić, L.; Chaput, D.L.; Haridas, S.; Wu, S.; LaButti, K.; Grigoriev, I.V.; Henrissat, B.; et al. Comparative Analysis of Secretome Profiles of Manganese(II)-Oxidizing Ascomycete Fungi. *PLoS ONE* **2016**, *11*, e0157844. [CrossRef]
62. Sharma, A.; Aggarwal, N.K.; Yadav, A. First Report of Lignin Peroxidase Production from *Alternaria alternata* ANF238 Isolated from Rotten Wood Sample. *Bioeng. Biosci.* **2016**, *4*, 76–87. [CrossRef]
63. Speight, J.G. *The Chemistry and Technology of Petroleum*; Marcel Decker: New York, NY, USA, 1999.
64. Hinkle, A.; Shin, E.-J.; Liberatore, M.W.; Herring, A.M.; Batzle, M. Correlating the chemical and physical properties of a set of heavy oils from around the world. *Fuel* **2008**, *87*, 3065–3070. [CrossRef]
65. Varjani, S.J. Microbial degradation of petroleum hydrocarbons. *Bioresour. Technol.* **2017**, *223*, 277–286. [CrossRef] [PubMed]
66. Van Hamme, J.D.; Singh, A.; Ward, O.P. Recent Advances in Petroleum Microbiology. *Microbiol. Mol. Biol. Rev.* **2003**, *67*, 503–549. [CrossRef] [PubMed]
67. Meulenberg, R.; Rijnaarts, H.H.; Doddema, H.J.; A Field, J. Partially oxidized polycyclic aromatic hydrocarbons show an increased bioavailability and biodegradability. *FEMS Microbiol. Lett.* **2006**, *152*, 45–49. [CrossRef]
68. Kulik, N.; Goi, A.; Trapido, M.; Tuhkanen, T. Degradation of polycyclic aromatic hydrocarbons by combined chemical pre-oxidation and bioremediation in creosote contaminated soil. *J. Environ. Manag.* **2006**, *78*, 382–391. [CrossRef] [PubMed]
69. Hamilton, P. Topographic Rossby waves in the Gulf of Mexico. *Prog. Oceanogr.* **2009**, *82*, 1–31. [CrossRef]
70. Portela, E.; Tenreiro, M.; Pallàs-Sanz, E.; Meunier, T.; Ruiz-Angulo, A.; Sosa-Gutiérrez, R.; Cusí, S. Hydrography of the central and western Gulf of Mexico. *J. Geophys. Res. Ocean.* **2018**, *123*, 5134–5149. [CrossRef]
71. Alboudwarej, H.; Felix, J.; Taylor, S.; Badry, R.; Bremner, C.; Brough, B.; Palmer, D.; Pattisson, K.; Beshry, M.; Krawchuk, P.; et al. Highlighting heavy oil. *Schlumberger Oilfield Rev. Summer* **2006**, *18*, 34–53.
72. Kingston, P. Long-term environmental impact of oil spills. *Splill Sci. Technol. Bull.* **2002**, *7*, 53–61. [CrossRef]
73. Atlas, R.M.; Hazen, T.C. Oil Biodegradation and Bioremediation: A Tale of the Two Worst Spills in U.S. History. *Environ. Sci. Technol.* **2011**, *45*, 6709–6715. [CrossRef] [PubMed]
74. Macaulay, B.M.; Rees, D. Bioremediation of oil spills: A review of challenges for research advancement. *Ann. Environ. Sci.* **2014**, *8*, 9–37.

Brief Report

The Presence of Marine Filamentous Fungi on a Copper-Based Antifouling Paint

Sergey Dobretsov [1,2,*], Hanaa Al-Shibli [3], Sajeewa S. N. Maharachchikumbura [4] and Abdullah M. Al-Sadi [3]

[1] Department of Marine Science and Fisheries, College of Agricultural and Marine Sciences, Sultan Qaboos University, P.O. Box 34, Al-Khod 123, Oman
[2] Centre of Excellence in Marine Biotechnology, Sultan Qaboos University, P.O. Box 50, Al-Khod 123, Oman
[3] Department of Crop Sciences, College of Agricultural and Marine Sciences, Sultan Qaboos University, P.O. Box 34, Al-Khod 123, Oman; Alshiblihana@gmail.com (H.A.-S.); alsadi@squ.edu.om (A.M.A.-S.)
[4] School of Life Science and Technology, University of Electronic Science and Technology of China, Chengdu 611731, China; sajeewa83@yahoo.com
* Correspondence: segey@squ.edu.om; Tel.: +968-2414-3581

Abstract: Marine biofouling is undesirable growth on submerged substances, which causes a major problem for maritime industries. Antifouling paints containing toxic compounds such as copper are used to prevent marine biofouling. However, bacteria and diatoms are usually found in biofilms developed on such paints. In this study, plastic panels painted with a copper-based self-polishing antifouling paint were exposed to biofouling for 6 months in the Marina Bandar Rowdha, Sea of Oman. Clean panels were used as a control substratum. Marine filamentous fungi from protected and unprotected substrate were isolated on a potato dextrose agar. Pure isolates were identified using sequences of the ITS region of rDNA. Six fungal isolates (*Alternaria* sp., *Aspergillus niger*, *A. terreus*, *A. tubingensis*, *Cladosporium halotolerans*, and *C. omanense*) were obtained from the antifouling paint. Four isolates (*Aspergillus pseudodeflectus*, *C. omanense*, and *Parengyodontium album*) were isolated from clean panels and nylon ropes. This is the first evidence of the presence of marine fungi on antifouling paints. In comparison with isolates from the unprotected substrate, fungi from the antifouling paint were highly resistant to copper, which suggests that filamentous fungi can grow on marine antifouling paints.

Keywords: marine fungi; copper; antifouling; coating; biofilm; Indian Ocean; Oman

Citation: Dobretsov, S.; Al-Shibli, H.; Maharachchikumbura, S.S.N.; Al-Sadi, A.M. The Presence of Marine Filamentous Fungi on a Copper-Based Antifouling Paint. *Appl. Sci.* **2021**, *11*, 8277. https://doi.org/10.3390/app11188277

Academic Editors: Anna Poli and Valeria Prigione

Received: 24 June 2021
Accepted: 20 August 2021
Published: 7 September 2021

1. Introduction

Marine biofouling is defined as the "undesirable accumulation and growth of organisms on submerged surfaces" [1]. Usually, biofouling organisms are divided by their size onto microfouling and macrofouling. Microfouling is composed of microscopic (<0.5 mm) organisms, mainly bacteria and diatoms [2,3]. Macrofouling, on the other hand, is composed of macroscopic organisms (>0.5 mm) visible by the naked eye, such as barnacles, mussels, bryozoans, macroalgae and others [4–6]. Microfouling has a significant impact on the recruitment of spores and larvae of algae and invertebrates (reviewed by [4,7]).

Marine biofouling causes significant problems for maritime industries [8,9]. It can increase the fuel consumption of ships, clog membranes and pipes, increase corrosion, decrease buoyancy, and destroy nets and cages [10,11]. Countries worldwide spend more than USD 7 billion per year in order to protect from biofouling and deal with is consequences [9].

In order to prevent submerged structures like boats and ships from biofouling companies are using antifouling coatings [9,12]. These antifouling coatings usually contain biocides that kill biofouling organisms. Currently, the most effective biocide is copper or cuprous oxide [12].

While antifouling paints are supposed to prevent biofouling, most of them have biofilms on their surfaces [3,13,14]. There is limited information about the biofilm composition of antifouling paints (see [15,16]). It has been shown that biofilms on antifouling paints consist of diverse species of bacteria and diatoms [17–19]. Up to now, filamentous fungi on antifouling paints have not been observed. Previous studies showed that biocides of antifouling paints and environmental conditions shaped the structure of the microbial communities [20,21].

Marine fungi are an important component of the marine environment [22,23]. While marine filamentous fungi are not very well studied, they are widely distributed and associated with sediments, sand grains, seaweeds, submerged wood, sea animals and plants [24]. Chytidiomycota and Ascomycota are fungal divisions dominated in samples from six European near-shore sites [22]. Forty-six fungal isolates belonging to the genera *Cladosporium, Paraphaeosphaeria, Trichoderma, Alternaria, Phoma*, and *Arthrinium* were isolated from marine biofilms developed on different submerged substrata [25]. Chytidiomycetes fungi dominated in marine biofilms developed on glass and plastic substrates [26]. To our knowledge, marine fungi have never been recorded on antifouling paints, especially copper-based ones. However, it was observed using metabarcoding that most of the fungal species in marine periphyton biofilms were not affected by 10 μM of copper [27]. Some species of marine filamentous fungi are highly resistant to high concentrations of copper. For example, *Penicillium chrysogenum* was able to tolerate concentrations of 500 mg L^{-1} of copper [28].

The main aim of this study was to identify the fungal isolates from biofilms developed on the surface of a cooper-based antifouling paint and demonstrate their copper resistance.

2. Materials and Methods

2.1. Antifouling Paint and Other Substrata

A commercial copper self-polishing paint Interspeed® BRA640 (International Paint, Gateshead, UK) was used in this study. The antifouling paint contains about 25–50% of cuprous oxide by weight [29]. An average release rate of copper from the paint was 3.8 μg cm^{-2} day^{-1} [30]. The paint was manually applied (thickness 125 μm) onto plastic fiberglass panels (15 cm × 28 cm) cleaned with ethanol (96%, Sigma, Ronkonkoma, NY, USA) in the laboratory. Fiberglass was obtained from a local Omani manufacturer (Al Kaboura, Muscat, Oman). This material was selected because it is used to make boats and it has a high biofouling potential. All coated panels were air-dried for several days at ambient temperature prior to deployment. All substrates (panels and ropes) were cleaned with 96% ethanol before the experiment to eliminate bacteria and fungi. No fungi were found on these substrates prior to the experiment.

2.2. Experiment and Testing Site

Three panels covered with the antifouling paint were exposed vertically to biofouling at the depth of 1 m for 6 months in Marina Bandar Rowdha (23,035′07″ N 58,036′48″ E), Muscat, Oman. As a control, uncoated fiberglass panels were used. Panels were fixed at the desired depth using a nylon rope (RopeNet, Taishan, China) attached to a pontoon. At the end of the rope, a weight was attached to keep the panels in a vertical position.

Marina Bandar Rowdha is a semi-enclosed bay for private recreational boats and yachts. It has a relatively high hydrocarbon and heavy metal pollution with one of the highest concentrations of TBT in Oman's waters [31]. This marina was selected due to its (very high) biofouling rates and a history of biofouling and antifouling investigations [21,32]. The experiment in the marina was conducted in 2018 between the months of February and September. During the study the seawater temperature varied from 24 to 30 °C, pH was about 8.2 and the salinity varied from 37 to 38 ppt. During the experiment, the seawater turbidity was 2–3 NTU (Nephelometric Turbidity Units).

2.3. Isolation of Fungi

Biofouled panels (painted and not) and ropes holding panels were collected from the marina in September 2018. At the marina, the ropes and the panels were individually packed into sterile bags and brought on ice to the laboratory. In the laboratory, the panels and ropes were washed several times with sterile distilled water (SDI). Using sterile scissors, the ropes were cut into 1.0 cm pieces. Surfaces of the panels and the ropes were disinfected using 1% sodium hypochlorite solution to eliminate bacteria (NaClO, Zhengzhou Sino Chemical Ltd., Beijing, China). Then, the panels and the ropes were washed three times with SDI. After that, the biofilms were removed from the panels using sterile cotton swabs. Finally, one piece from each rope or an individual swab was placed into a Petri dish containing a 2.5% potato dextrose agar (PDA, Merck, Kenilworth, NJ, USA) prepared using filtered (0.45 μm cellulose nitrate filter, Sartorius, Germany) and autoclaved seawater from the marina. As a control, PDA Petri dishes containing autoclaved seawater were used. Visible growth of fungi was checked after incubation at 25 °C for up to three weeks. Each individual fungal colony was transferred into a new fresh PDA plate. Pure fungal colonies were stored on PDA slants with 10% glycerol for further genetic identification (see below).

2.4. Identification of Fungi

Before the identification, the isolate was grown on PDA. The identification of filamentous fungal isolates was done based on sequences of the internal transcribed spacer region (ITS) of the ribosomal DNA [33]. Firstly, 80 g of fungal mycelia were harvested and freeze-dried. Then, its DNA was further extracted [34]. Secondly, the ITS rDNA region was amplified using the primer pairs of ITS4 (TCCTCCGCTTATTGATATGC) and ITS5 (GGAAGTAAAAGT CGTAACAAGG). The PCR program followed the conditions of [35]. MACROGEN, Korea sequenced the PCR products. In order to obtain the ITS rRNA sequence, two complementary sequences for each fungal isolate were aligned using MEGA v.6 [36]. Fungal isolates were identified based on a comparison of the ITS rRNA sequences against the National Center for Biotechnology Information (NCBI) database. Sequences of fungal isolates were deposited in the NCBI GenBank database with accession numbers MN947598–MN947607. For phylogenetic trees, maximum likelihood analysis with 1000 bootstrap replicates based on ITS sequence data was done using RaxmlGUI v. 1.3 [37]. The final phylogenetic tree was selected by comparing the likelihood scores using the GTR+GAMMA substitution model.

2.5. Copper Resistance of Fungal Isolates

In order to prove that fungal isolates are able to grow on antifouling paints, we tested their sensitivity to copper by an agar diffusion technique. Because copper oxide is not soluble in water, copper sulfate (CuSO4, Sigma Aldrich, Ronkonkoma, NY, USA) was used. Firstly, different concentrations (500–0.01 g L^{-1}) of copper sulfate in autoclaved seawater were prepared. Secondly, fungal isolates were grown onto 2.5% potato dextrose agar (PDA, Merck, Kenilworth, NJ, USA) for 3 days. PDA was made using autoclaved seawater from the marina. A four mm disc of each fungal isolate was cut and individually placed onto the PDA Petri dish. Thirdly, 10 μL of copper sulfate solution was added to a sterile paper disk (diameter 6 mm). As a control, disks with 10 μL of seawater were used. The disks were air-dried at room temperature and placed in the middle of a PDA Petri dish two cm away from the isolate. The dishes were incubated at 25 °C for 5 days. The experiment was made in triplicate. The presence or absence of an inhibition zone was detected. Finally, the minimal inhibitory concentration of copper (II) sulfate (μg cm^{-2}) for each fungal isolate was calculated.

3. Results and Discussion

3.1. Biofouling on Different Substrata

Biofouling on the antifouling paint was minimal and only biofilms were observed. In opposite, the ropes and unprotected panels were completely covered with macrofouling organisms, dominated by Tunicata and Bryozoa. This supports our previous data about performance of different antifouling paints in Oman waters [21,32]. The 1-year field experiment showed that copper-based antifouling paints have only diatom and bacterial biofouling [21]. A previous experiment with unprotected fiberglass and acrylic panels demonstrated dominance of Bryozoa, barnacles and sponges [38]. The absence of sponges and barnacles in the current study could be due to differences in the substratum chemistry (nylon versus acrylic) and shape (flat plates versus cylindrical ropes).

3.2. Species of Fungi Isolated from Different Substrata

In total, six fungal isolates were obtained from the antifouling paint and four were isolated from unprotected substrata (Table 1). Based on the phylogenetic analysis, the majority of isolates belonged to *Aspergillus* and *Cladosporium* genera (Supplement Figures S1–S4). The genera *Aspergillus* and *Cladosporium* are commonly found in the marine environment [39–41]. Additionally, *Aspergillus* and *Cladosporium* are associated with marine sponges [42,43]. There is limited information about marine derived filamentous fungi in Oman, but we have been able to isolate *Aspergillus terreus* from mangrove areas [44]. *Cladosporium omanense* found in this study (Table 1) was previously isolated from living leaves of *Zygophyllum coccineum* in Oman [45]. The presence of *C. omanense* on all investigated substrates could be due to several reasons. It could suggest that this species is very common in Omani waters and can colonize protected and unprotected substrata. Alternatively, it could be due to contamination of our culture by spores of this fungus. This is highly unlikely, as there were no fungi recovered from the control plates with autoclaved seawater.

Filamentous fungi belonging to the genera *Parengyodontium* were isolated from biofouled ropes only (Table 1). *Parengyodontium album* is an environmental saprobic mold and an opportunistic pathogen [46]. This species has been observed on buildings composed of limestone and plaster [47]. Additionally, *P. album* was found in sediments of polar-boreal White Sea [48]. The presence of this fungus in Oman waters suggests that this species can be found in tropical waters as well.

The genera *Aspergillus, Cladosporium* and *Alternaria* were found on the copper-based antifouling paint (Table 1). Moreover, *A. tubingensis, A. terreus, A. niger* and *C. halotolerans* were found only on the antifouling paint. *Alternaria* isolates were obtained exclusively from the paint. Previously, the fungi *Alternaria* were isolated from soft corals [49], sponges [50] and algae [51]. While 18S RNA of fungi belonging to the class Agaricomycetes was detected on an antifouling paint using Illumina amplicon sequencing [52], fungal isolates were obtained from antifouling paints for the first time in this study. Previously, only bacteria and diatoms were detected in biofilms on antifouling paints [3,21].

3.3. Copper Resistance of Fungal Isolates

In order to prove that fungal isolates are able to grow on antifouling paints, their sensitivity to different copper concentrations is tested in laboratory experiments (Table 2). Due to low solubility of CuO, $CuSO_4$ was used in this experiment. Previous studies suggest that $CuSO_4$ is more toxic compare to CuO [53]. Thus, the isolates are more resistant to CuO than is reported in Table 2. Generally, isolates from antifouling paint can tolerate higher concentrations of copper. Five out of six isolates from the antifouling paint can tolerate an average daily release rate of copper 3.8 μg cm^{-2} day^{-1} [30] from the tested paint (Table 2). The highest copper resistance was observed for *Aspergillus terreus*. This fungus can tolerate 2% of $CuCl_2$ in a polyvinyl chloride coating in a laboratory experiment [54] and can be used to remove heavy metals from water [54]. In opposite, fungal isolates from unprotected substrata had low tolerance to copper (Table 2). This suggests that isolates from the copper-based antifouling paint are adapted to high copper concentrations and could grow and play an important role in biofilms.

Table 1. The list of fungal isolates from panels painted with the antifouling paint and not painted (control) panels and ropes.

Species	Substrate	No	GenBank Accession Number	Picture
Aspergillus tubingensis	Antifouling paint	H1	MN947598	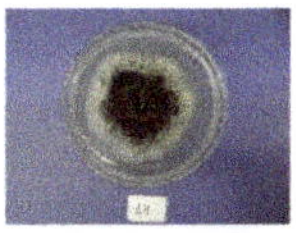
Aspergillus terreus	Antifouling paint	H2	MN947599	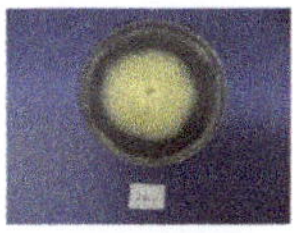
Alternaria sp.	Antifouling paint	H3	MN947600	
Aspergillus niger	Antifouling paint	H4	MN947601	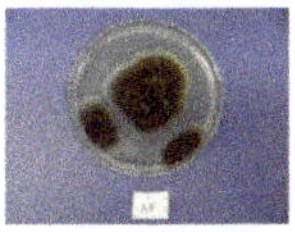
Cladosporium halotolerans	Antifouling paint	H6	MN947602	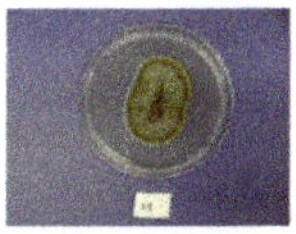
Cladosporium omanense	Antifouling paint	H7	MN947603	
Aspergillus pseudodeflectus	Not painted panel (control)	H90	MN947605	
Cladosporium omanense	Not painted panel (control)	H89	MN947604	
Cladosporium omanense	Ropes	H91	MN947606	
Parengyodontium album	Ropes	H92	MN947607	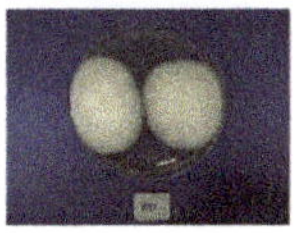

Table 2. The minimal inhibitory concentration of copper (II) sulfate (µg cm^{-2}) for fungal isolates from panels painted with the antifouling paint, not painted (control), and ropes. Highlighted values exceed an average release rate of copper from the paint (3.8 µg cm^{-2} day^{-1} [30]).

Species	Substrate	Minimal Inhibitory Concentration
Aspergillus tubingensis	Antifouling paint	**4.3**
Aspergillus terreus	Antifouling paint	**5.2**
Alternaria sp.	Antifouling paint	**3.9**
Aspergillus niger	Antifouling paint	**4.3**
Cladosporium halotolerans	Antifouling paint	**3.9**
Cladosporium omanense	Antifouling paint	1.3
Aspergillus pseudodeflectus	Control	0.17
Cladosporium omanense	Control	0.87
Cladosporium omanense	Ropes	1.3
Parengyodontium album	Ropes	1.3

3.4. Importance of This Study

Our finding has very important implications for antifouling industries. Firstly, it demonstrates that some fungal species can live on antifouling paints and tolerate relatively high copper concentrations. Compared to the isolates from unprotected substrata (ropes and panels), fungi from the antifouling paint were highly resistant to copper. Previous studies suggested copper resistance of some fungal species that bind copper to cell walls [55,56]. Additionally, fungi can produce copper-binding proteins and chelating compounds in response to elevated concentrations of copper [55]. Secondly, the role of fungal species on antifouling paints requires further investigations. It is possible to propose that filamentous fungi can degrade organic matrix of the paint, composed of vinyl or acrylic resin or silicone polymers, which in turn can affect release of the biocide and the life span of the antifouling paint. It has been shown that filamentous fungi can deteriorate synthetic paints [57,58]. Additionally, filamentous fungi can degrade biocides of antifouling paints, such as Irgarol 1051 [59] and TBT [60]. In opposite, some marine fungal species produce antimicrobial and antifouling compounds (see review [61]). Presence of these strains on paints could be beneficial and enhance their antifouling properties. Finally, more research is needed to understand if marine fungi can be found on other antifouling paints exposed to biofouling in different seas and investigate possible mechanisms whereby fungi transform these paints. Additionally, it is important to investigate the role of marine fungi on antifouling paints and possible mechanisms of their resistance.

Supplementary Materials: The following are available online at https://www.mdpi.com/article/10.3390/app11188277/s1, Figure S1: Phylogram generated from maximum likelihood analysis based on ITS sequence data of analyzed *Alternaria* species. Isolates derived from this study are in red. The tree is rooted to *A. alternantherae* (CBS124392), Figure S2: Phylogram generated from maximum likelihood analysis based on ITS sequence data of analyzed *Aspergillus* species. Isolates derived from this study are in red. The tree is rooted to *Penicillium herquei* (CBS 336.48), Figure S3: Phylogram generated from maximum likelihood analysis based on ITS sequence data of analyzed *Cladosporium* species. Isolates derived from this study are in red, Figure S4: The tree is rooted to *Cercospora beticola* (CBS 116456), Phylogram generated from maximum likelihood analysis based on ITS sequence data of analyzed *Parengyodontium* species. Isolates derived from this study are in red. The tree is rooted to *Purpureocillium lilacinum* (CBS 284.36).

Author Contributions: Conceptualization, S.D. and A.M.A.-S.; methodology, S.D. and A.M.A.-S.; formal analysis, H.A.-S., S.D. and S.S.N.M.; investigation, H.A.-S. and S.S.N.M.; writing—original draft preparation, S.D. and A.M.A.-S.; writing—review and editing, S.D., A.M.A.-S., H.A.-S. and S.S.N.M.; supervision, S.D. and A.M.A.-S.; project administration, S.D. and A.M.A.-S.; funding acquisition, S.D. and A.M.A.-S. All authors have read and agreed to the published version of the manuscript.

Funding: This research was funded by the TRC grant RC/AGR/FISH/16/01, Omantel grant EG/SQU-OT/20/01, SQU internal grant IG/AGR/FISH/18/01 and the grant EG/AGR/CROP/16/01.

Institutional Review Board Statement: Not applicable.

Informed Consent Statement: Not applicable.

Data Availability Statement: Sequences of fungal isolates can be found in the NCBI GenBank database with accession numbers MN947598–MN947607.

Acknowledgments: S.D. acknowledged financial support by the TRC grant RC/AGR/FISH/16/01 and SQU internal grant IG/AGR/FISH/18/01. A.M.A. research was supported by the grant EG/AGR/CROP/16/01 and RC/AGR/FISH/16/01.

Conflicts of Interest: The authors declare no conflict of interest.

References

1. Wahl, M. Marine epibiosis. I. Fouling and antifouling: Some basic aspects. *Mar. Ecol. Prog. Ser.* **1989**, *58*, 175–189. [CrossRef]
2. Dobretsov, S.; Dahms, H.-U.; Qian, P.-Y. Inhibition of biofouling by marine microorganisms and their metabolites. *Biofouling* **2006**, *22*, 43–54. [CrossRef] [PubMed]
3. Salta, M.; Wharton, J.A.; Blache, Y.; Stokes, K.R.; Briand, J.-F. Marine biofilms on artificial surfaces: Structure and dynamics. *Environ. Microbiol.* **2013**, *15*, 2879–2893. [CrossRef] [PubMed]
4. Hadfield, M.G. Biofilms and marine invertebrate larvae: What bacteria produce that larvae use to choose settlement sites. *Annu. Rev. Mar. Sci.* **2011**, *3*, 453–470. [CrossRef]
5. Venkatesan, R.; Murthy, P.S. Macrofouling control in power plants. In *Marine and Industrial Biofouling*; Flemming, H.C., Murthy, P.S., Venkatesan, R., Cooksey, K., Eds.; Springer Series on Biofilms; Springer: Berlin/Heidelberg, Germany, 2009; Volume 4, pp. 1–27.
6. Wieczorek, S.K.; Todd, C.D. Inhibition and facilitation of settlement of epifaunal marine invertebrate larvae by microbial biofilm cues. *Biofouling* **1998**, *12*, 81–118. [CrossRef]
7. Dobretsov, S.; Rittschof, D. Love at first taste: Induction of larval settlement by marine microbes. *Int. J. Mol. Sci.* **2020**, *21*, 731. [CrossRef]
8. Schultz, M.P.; Bendick, J.A.; Holm, E.R.; Hertel, W.M. Economic impact of biofouling on a naval surface ship. *Biofouling* **2011**, *27*, 87–98. [CrossRef]
9. Yebra, D.M.; Kiil, S.; Dam-Johansen, K. Antifouling technology—Past, present and future steps towards efficient and environmentally friendly antifouling coatings. *Prog. Org. Coat.* **2004**, *50*, 75–104. [CrossRef]
10. Bott, T.R. Biofouling control in cooling water. *Int. J. Chem. Eng.* **2009**, *2009*. [CrossRef]
11. Jones, G. The battle against marine biofouling: A historical review. In *Advances in Marine Antifouling Coatings and Technologies*; Hellio, C., Yebra, D., Eds.; Woodhead publishing series in metals and surface engineering; CRC Press: Boca Ration, FL, USA, 2009; pp. 19–45.
12. Maréchal, J.-P.; Hellio, C. Challenges for the development of new non-toxic antifouling solutions. *Int. J. Mol. Sci.* **2009**, *10*, 4623–4637. [CrossRef]
13. Dempsey, M.J. Colonisation of antifouling paints by marine bacteria. *Bot. Mar.* **2009**, *24*, 185–192. [CrossRef]
14. Thomas, T.E.; Robinson, M.G. The role of bacteria in the metal tolerance of the fouling diatom *Amphora coffeaeformis* Ag. *J. Exp. Mar. Biol. Ecol.* **1987**, *107*, 291–297. [CrossRef]
15. Yang, J.-L.; Li, Y.-F.; Liang, X.; Guo, X.-P.; Ding, D.W.; Zhang, D.; Zhou, S.; Bao, W.-Y.; Bellou, N.; Dobretsov, S. Silver nanoparticles impact biofilm communities and mussel settlement. *Sci. Rep.* **2016**, *6*, 37406. [CrossRef]
16. Yang, J.-L.; Li, Y.-F.; Guo, X.-P.; Liang, X.; Xu, Y.-F.; Ding, D.-W.; Bao, W.-Y.; Dobretsov, S. The effect of carbon nanotubes and titanium dioxide incorporated in PDMS on biofilm community composition and subsequent mussel plantigrade settlement. *Biofouling* **2016**, *32*, 763–777. [CrossRef]
17. Chen, C.-L.; Maki, J.S.; Rittschof, D.; Teo, S.L.-M. Early marine bacterial biofilm on a copper-based antifouling paint. *Int. Biodeterior. Biodegrad.* **2013**, *83*, 71–76. [CrossRef]
18. Pelletier, É.; Bonnet, C.; Lemarchand, K. Biofouling growth in cold estuarine waters and evaluation of some chitosan and copper anti-fouling paints. *Int. J. Mol. Sci.* **2009**, *10*, 3209–3223. [CrossRef]
19. Winfield, M.O.; Downer, A.; Longyer, J.; Dale, M.; Barker, G.L.A. Comparative study of biofilm formation on biocidal antifouling and fouling-release coatings using next-generation DNA sequencing. *Biofouling* **2018**, *34*, 464–477. [CrossRef]
20. Briand, J.-F.; Djeridi, I.; Jamet, D.; Cope, S.; Bressy, C.; Molmeret, M.; Le Berre, B.; Rimet, F.; Bouchez, A.; Blache, Y. Pioneer marine biofilms on artificial surfaces including antifouling coatings immersed in two contrasting French Mediterranean coast sites. *Biofouling* **2012**, *28*, 453–463. [CrossRef] [PubMed]
21. Muthukrishnan, T.; Abed, R.M.M.; Dobretsov, S.; Kidd, B.; Finnie, A.A. Long-term microfouling on commercial biocidal fouling control coatings. *Biofouling* **2014**, *30*, 1155–1164. [CrossRef]

22. Richards, T.A.; Leonard, G.; Mahé, F.; del Campo, J.; Romac, S.; Jones, M.D.M.; Maguire, F.; Dunthorn, M.; De Vargas, C.; Massana, R.; et al. Molecular diversity and distribution of marine fungi across 130 European environmental samples. *Proc. R. Soc. B Biol. Sci.* **2015**, *282*, 20152243. [CrossRef] [PubMed]
23. Tisthammer, K.H.; Cobian, G.M.; Amend, A.S. Global biogeography of marine fungi is shaped by the environment. *Fung. Ecol.* **2016**, *19*, 39–46. [CrossRef]
24. Overy, P.D.; Rämä, T.; Oosterhuis, R.; Walker, K.A.; Pang, K.-L. The neglected marine fungi, *sensu stricto*, and their isolation for natural products' discovery. *Mar. Drug.* **2019**, *17*, 42. [CrossRef] [PubMed]
25. Miao, L.; Qian, P.-Y. Antagonistic antimicrobial activity of marine fungi and bacteria isolated from marine biofilm and seawaters of Hong Kong. *Aquat. Microb. Ecol.* **2005**, *38*, 231–238. [CrossRef]
26. Kirstein, I.V.; Wichels, A.; Krohne, G.; Gerdts, G. Mature biofilm communities on synthetic polymers in seawater-specific or general? *Mar. Environ. Res.* **2018**, *142*, 147–154. [CrossRef] [PubMed]
27. Corcoll, N.; Yang, J.; Backhaus, T.; Zhang, X.; Eriksson, K.M. Copper affects composition and functioning of microbial communities in marine biofilms at environmentally relevant concentrations. *Front. Microb.* **2019**, *9*, 3248. [CrossRef]
28. Lotlikar, N.; Damare, S.; Meena, R.M.; Jayachandran, S. Variable protein expression in marine-derived filamentous fungus *Penicillium chrysogenum* in response to varying copper concentrations and salinity. *Metallomics* **2020**, *12*, 1083–1093. [CrossRef] [PubMed]
29. Interspeed BRA640 Red Safety Sheet. Available online: http://datasheets1.international-coatings.com/msds/BRA640_USA_eng_B4.pdf (accessed on 22 July 2020).
30. Valkirs, A.O.; Seligman, P.F.; Haslbeck, E.; Caso, J.S. Measurement of copper release rates from antifouling paint under laboratory and in situ conditions: Implications for loading estimation to marine water bodies. *Mar. Poll. Bull.* **2003**, *46*, 763–779. [CrossRef]
31. Jupp, B.P.; Fowler, S.W.; Dobretsov, S.; van der Wiele, H.; Al-Ghafri, A. Assessment of heavy metal and petroleum hydrocarbon contamination in the Sultanate of Oman with emphasis on harbours, marinas, terminals and ports. *Mar. Pollut. Bull.* **2017**, *121*, 260–273. [CrossRef]
32. Muthukrishnan, T.; Dobretsov, S.; De Stefano, M.; Abed, R.M.M.; Kidd, B.; Finnie, A.A. Diatom communities on commercial biocidal fouling control coatings after one year of immersion in the marine environment. *Mar. Environ. Res.* **2017**, *129*, 102–112. [CrossRef]
33. Al-Sadi, A.M.; Al-Mazroui, S.S.; Phillips, A.J.L. Evaluation of culture-based techniques and 454 pyrosequencing for the analysis of fungal diversity in potting media and organic fertilizers. *J. Appl. Microb.* **2015**, *119*, 500–509. [CrossRef]
34. Al-Sadi, A.M.; Al-Ghaithi, A.G.; Al-Balushi, Z.M.; Al-Jabri, A.H. Analysis of diversity in *Pythium aphanidermatum* populations from a single greenhouse reveals phenotypic and genotypic changes over 2006 to 2011. *Plant Dis.* **2012**, *96*, 852–858. [CrossRef]
35. White, T.J.; Bruns, T.D.; Lee, S.; Taylor, J. Amplification and direct sequencing of fungal ribosomal RNA genes for phylogenetics. In *PCR Protocols: A Guide to Methods and Applications*; Innis, M.A., Gelfand, D.H., Sninsky, J.J., White, T.J., Eds.; Academic Press: New York, NY, USA, 1990; pp. 315–322.
36. Tamura, K.; Stecher, G.; Peterson, D.; Filipski, A.; Kumar, S. MEGA6: Molecular evolutionary genetics analysis version 6.0. *Mol. Biol. Evol.* **2013**, *30*, 2725–2729. [CrossRef]
37. Silvestro, D.; Michalak, I. RaxmlGUI: A graphical front-end for RAxML. *Org. Divers. Evol.* **2012**, *12*, 335–337. [CrossRef]
38. Dobretsov, S. Biofouling on artificial substrata in Muscat waters. *J. Agric. Mar. Sci. JAMS* **2015**, *20*, 24–29. [CrossRef]
39. Chung, D.; Kim, H.; Choi, H.S. Fungi in salterns. *J. Microbiol. Seoul Korea* **2019**, *57*, 717–724. [CrossRef]
40. Corinaldesi, C.; Barone, G.; Marcellini, F.; Dell'Anno, A.; Danovaro, R. Marine microbial-derived molecules and their potential use in cosmeceutical and cosmetic products. *Mar. Drugs* **2017**, *15*, 118. [CrossRef]
41. Lee, Y.M.; Kim, M.J.; Li, H.; Zhang, P.; Bao, B.; Lee, K.J.; Jung, J.H. Marine-derived *Aspergillus* species as a source of bioactive secondary metabolites. *Mar. Biotechnol.* **2013**, *15*, 499–519. [CrossRef]
42. Paulino, G.V.B.; Félix, C.R.; Landell, M.F. Diversity of filamentous fungi associated with coral and sponges in coastal reefs of northeast Brazil. *J. Basic Microbiol.* **2019**, *60*, 103–111. [CrossRef]
43. Wang, G.; Li, Q.; Zhu, P. Phylogenetic diversity of culturable fungi associated with the Hawaiian sponges *Suberites zeteki* and *Gelliodes fibrosa*. *Antonie Van Leeuwenhoek* **2008**, *93*, 163–174. [CrossRef] [PubMed]
44. Al-Shibli, H.; Dobretsov, S.; Al-Nabhani, A.; Maharachchikumbura, S.S.N.; Rethinasamy, V.; Al-Sadi, A.M. *Aspergillus terreus* obtained from mangrove exhibits antagonistic activities against Pythium aphanidermatum-induced damping-off of cucumber. *PeerJ* **2019**, *7*, e7884. [CrossRef]
45. Halo, B.A.; Maharachchikumbura, S.S.N.; Al-Yahyai, R.A.; Al-Sadi, A.M. *Cladosporium omanense*, a new endophytic species from *Zygophyllum coccineum* in Oman. *Phytotaxa* **2019**, *388*, 145–154. [CrossRef]
46. Tsang, C.-C.; Chan, J.F.W.; Pong, W.-M.; Chen, J.H.K.; Ngan, A.H.Y.; Cheung, M.; Lai, C.K.C.; Tsang, D.N.C.; Lau, S.K.P.; Woo, P.C.Y. Cutaneous hyalohyphomycosis due to *Parengyodontium album* gen. et comb. nov. *Med. Mycol.* **2016**, *54*, 699–713. [CrossRef]
47. Ponizovskaya, V.B.; Rebrikova, N.L.; Kachalkin, A.V.; Antropova, A.B.; Bilanenko, E.N.; Mokeeva, V.L. Micromycetes as colonizers of mineral building materials in historic monuments and museums. *Fungal Biol.* **2019**, *123*, 290–306. [CrossRef]
48. Khusnullina, A.I.; Bilanenko, E.N.; Kurakov, A.V. Microscopic fungi of White Sea sediments. Contemp. *Probl. Ecol.* **2018**, *11*, 503–513. [CrossRef]

49. Zhao, D.-L.; Cao, F.; Wang, C.-Y.; Yang, L.-J.; Shi, T.; Wang, K.-L.; Shao, C.-L.; Wang, C.-Y. Alternatone A, an unusual perylenequinone-related compound from a soft-coral-derived strain of the fungus *Alternaria alternata*. *J. Nat. Prod.* **2019**, *82*, 3201–3204. [CrossRef]
50. Chen, Y.; Chen, R.; Xu, J.; Tian, Y.; Xu, J.; Liu, Y. Two new altenusin/thiazole hybrids and a new benzothiazole derivative from the marine sponge-derived fungus *Alternaria* sp. SCSIOS02F49. *Molecules* **2018**, *23*, 2844. [CrossRef]
51. Shi, Z.-Z.; Fang, S.-T.; Miao, F.-P.; Ji, N.-Y. Two new tricycloalternarene esters from an alga-epiphytic isolate of *Alternaria alternata*. *Nat. Prod. Res.* **2018**, *32*, 2523–2528. [CrossRef]
52. Dobretsov, S.; Abed, R.M.M.; Muthukrishnan, T.; Sathe, P.; Al-Naamani, L.; Queste, B.Q.; Piontkovski, S. Living on the edge: Biofilms developing in oscillating environmental conditions. *Biofouling* **2018**, *34*, 1064–1077. [CrossRef]
53. Rotini, A.; Tornambè, A.; Cossi, R.; Iamunno, F.; Benvenuto, G.; Berducci, M.T.; Maggi, C.; Thaller, M.C.; Cicero, A.M.; Manfra, L.; et al. Salinity-based toxicity of CuO nanoparticles, CuO-bulk and Cu ion to *Vibrio anguillarum*. *Front. Microbiol.* **2017**, *8*, 2076. [CrossRef]
54. Nasrallah, D.A.; Morsi, M.A.; El-Sayed, F.; Metwally, R.A. Structural, optical and electrical properties of copper chloride filled polyvinyl chloride/polystyrene blend and its antifungal properties against *Aspergillus avenaceus* and *Aspergillus terreus*. *Compos. Commun.* **2020**, *22*. [CrossRef]
55. Dias, M.A.; Lacerda, I.C.A.; Pimentel, P.F.; De Castro, H.F.; Rosa, C.A. Removal of heavy metals by an *Aspergillus terreus* strain immobilized in a polyurethane matrix. *Lett. Appl. Microb.* **2002**, *34*, 46–50. [CrossRef]
56. Cervantes, C.; Gutierrez-Corona, F. Copper resistance mechanisms in bacteria and fungi. *FEMS Microbiol. Rev.* **1994**, *14*, 121–137. [CrossRef]
57. Cappitelli, F.; Vicini, S.; Piaggio, P.; Abbruscato, P.; Princi, E.; Casadevall, A.; Nosanchuk, J.D.; Zanardini, E. Investigation of fungal deterioration of synthetic paint binders using vibrational spectroscopic techniques. *Macromol. Biosci.* **2005**, *5*, 49–57. [CrossRef]
58. Saad, D.S.; Kinsey, G.C.; Kim, S.; Gaylarde, C.C. Extraction of genomic DNA from filamentous fungi in biofilms on water-based paint coatings; 12th International Biodeterioration and Biodegradation Symposium (Biosorption and Bioremediation III). *Int. Biodeterior. Biodegrad.* **2004**, *54*, 99–103. [CrossRef]
59. Ogawa, N.; Okamura, H.; Hirai, H.; Nishida, T. Degradation of the antifouling compound Irgarol 1051 by manganese peroxidase from the white rot fungus *Phanerochaete chrysosporium*. *Chemosphere* **2004**, *55*, 487–491. [CrossRef]
60. Bernat, P.; Długoński, J. Acceleration of tributyltin chloride (TBT) degradation in liquid cultures of the filamentous fungus *Cunninghamella elegans*. *Chemosphere* **2006**, *62*, 3–8. [CrossRef]
61. Liu, L.L.; Wu, C.-H.; Qian, P.Y. Marine natural products as antifouling molecules—A mini-review (2014–2020). *Biofouling* **2020**, *36*, 1210–1226.

applied sciences

MDPI

Review

Facing Phototrophic Microorganisms That Colonize Artistic Fountains and Other Wet Stone Surfaces: Identification Keys

Fernando Bolivar-Galiano [1], Oana Adriana Cuzman [2,3,*], Clara Abad-Ruiz [1] and Pedro Sánchez-Castillo [4]

1 Department of Painting, Faculty of Fine Arts, University of Granada, 18071 Granada, Spain; fbolivar@ugr.es (F.B.-G.); cabad@ugr.es (C.A.-R.)
2 Institute of Heritage Science-National Research Council (ISPC-CNR), 50019 Florence, Italy
3 Opificio delle Pietre Dure (OPD), Via degli Alfani 78, 50121 Florence, Italy
4 Department of Vegetal Biology, Faculty of Sciences, University of Granada, 18071 Granada, Spain; psanchez@ugr.es
* Correspondence: oanaadriana.cuzman@cnr.it; Tel.: +39-055-522-5479

Featured Application: This work presents a helpful tool when dealing with the biodiversity of monumental fountains and other wet lithotype surfaces in the sector of stone conservation.

Abstract: All fountains are inhabited by phototrophic microorganisms, especially if they are functional and located outdoors. This fact, along with the regular presence of water and the intrinsic bioreceptivity of stone material, easily favors the biological development. Many of these organisms are responsible for the biodeterioration phenomena and recognizing them could help to define the best strategies for the conservation and maintenance of monumental fountains. The presence of biological growth involves different activities for the conservation of artistic fountains. This paper is a review of the phototrophic biodiversity reported in 46 fountains and gives a whole vision on coping with biodeteriogens of fountains, being an elementary guide for professionals in the field of stone conservation. It is focused on recognizing the main phototrophs by using simplified dichotomous keys for cyanobacteria, green algae and diatoms. Some basic issues related to the handling of the samples and with the control of these types of microalgae are also briefly described, in order to assist interested professionals when dealing with the biodiversity of monumental fountains.

Keywords: green algae; diatoms; cyanobacteria; microalgae; stone conservation; diagnosis tool; preservation strategies; biodeterioration

Citation: Bolivar-Galiano, F.; Cuzman, O.A.; Abad-Ruiz, C.; Sánchez-Castillo, P. Facing Phototrophic Microorganisms That Colonize Artistic Fountains and Other Wet Stone Surfaces: Identification Keys. *Appl. Sci.* **2021**, *11*, 8787. https://doi.org/10.3390/app11188787

Academic Editors: Cédric Delattre, Anna Poli and Valeria Prigione

Received: 9 July 2021
Accepted: 13 September 2021
Published: 21 September 2021

1. Introduction

Fountains are structures with a functional, decorative or recreation purpose that have been built since ancient times [1]. Monumental and artistic fountains are often impressive constructions built all over the world, belonging to the public and garden art, such as 'La Joute' Fountain from Montreal (Canada), Fountain of Giant Wild Goose Pagoda from Xi'an (China), Flora Fountain from Mumbai (India), Keller Fountain from Portland, Oregon (USA), etc. They are mainly made of stone (structure, basin and decorations) and metallic alloys (hydraulic system and decorations). Stones with a local origin are preferred, but valuable stones are also used for the significant decorative elements (e.g., marble). However, other types of materials have been employed in the recent century, such as concrete material (e.g., Villancourt Fountain, San Francisco, CA, USA) or other modern materials such resins (Stravinsky Fountain, Paris, France, or Charybdis' Vortex Fountain, Sunderland, UK).

The roughness of the stone substrate is one of the main factors that influences the phototrophic colonization by epilithic, chasmolithic or endolithic microorganisms [2]. The last ones include the criptoendolithic microorganisms (which develop in layers, parallel with respect to the surface) and euendolithic microorganisms (which are the most aggressive, as they actively penetrate inside the stone). The epilithic communities are the most frequent. They mainly induce esthetical damage, but often corrosion phenomena can be observed (like little holes in the stones, corresponding to the phototrophic settlement). The chasmolithic development induces damage such as detachments or stone flakes formation and lifting. The biodeterioration phenomena were investigated in different studies [3–7].

The main types of phototrophic communities form pellicles, pustules, mats or layered sedimentary formations [6]. Often, the biological presence of phototrophic microorganisms can be considered a patina [8] or a phototrophic biofilm (due to its mucilaginous aspect).

Artistic fountains are not only heritage items but sociocultural artefacts as well, with a significant importance for the identity of certain places or periods [9,10]. They are part of the community and human treasures and raise preservation demands for transmitting their values to the further generations. Applying monitoring activity with a professional maintenance plan is essential for good and effective preservation of these structures. The preservation strategies should include different activities such as cleaning, repairs and consolidation, application or replacement of protective coatings, controlling of the water quality; all these interventions are related with the degradation problems induced by physical, chemical or biological agents. Coping with the phototrophic colonizers of artistic fountains and other wet surfaces means mainly cleaning and monitoring activity. The morphological identification of the phototrophic presence is usually enough for the documentation before and after the restoration. A detailed mapping of all types of deterioration phenomena present on a certain case study helps to decide the best strategies for both treatment and monitoring plans. Although the control treatments are often common for all the phototrophic groups, it worth mentioning that the principle of minimal intervention should always be kept in mind, being therefore ready to undertake, when necessary, selective treatments on the same stone artefact. This is due especially to two main reasons: (i) the phototrophic growth on the fountains is rarely formed by a singular group of phototrophs, and (ii) the cleaning treatments (chemical, physical or mechanical) are not so selective. However, in particular cases (e.g., the presence of phototrophic endolithic growth) it is possible to improve the cleaning procedures by making changes to the applications time, the concentration used, or to combine different types of cleaning treatments (e.g., mechanical and chemical; mechanical and physical). Moreover, the molecular identification could be necessary despite increasing procedure costs. These types of analysis need a small amount of sample and are definitely more precise than the morphological analysis. They are very useful to control the regrowth of dangerous species after a certain treatment.

An easy to use guide containing identification keys for the main phototrophs growing on monumental fountains and other wet stone surfaces is presented in this work. These new keys resulted after revising and simplifying other different identification keys [11–21] and they were tailored as a new tool for professionals in the field of stone conservation, that are interested in making their first steps in phototrophs identification. It is based on both macroscopic and microscopic characteristics for the main genera belonging to cyanobacteria, green algae and diatoms, the main three phototrophic groups found on the ornamental fountains. Moreover, some basic issues related to the handling of the samples and the control methodologies were also briefly described.

2. Materials and Methods

2.1. Morphological Identification

Phototrophic samples (about 265) were collected from various ornamental fountains and identified by the authors of this paper in precedent works [21–26] as indicated in Tables 1–4. The most frequent genera found by other authors [3,5,7,24,25,27–32] are also inserted in these identification keys.

Table 1. Artistic fountains investigated for the main phototrophic microorganisms.

Location	Fountains List
A = San Agustín-Huila, Colombia [5]	**1** = *Fountain of Lavapatas* (green area, made of volcanic rock)
B = Lamezia Terme, Italy [3]	**2** = *New Fountain*
C = Florence, Italy [23]	**3** = *Tacca's Fountain* (urban area, made of marble (pedestal and border of basins) and pietra Serena sandstone (basins))
	4 = *Second Fountain* of Villa la Pietra (green area, made of concrete)
D = Palermo, Italy [7]	**5** = *Pretoria Fountain* (urban area, made of marble (decorative elements and border of the basin), red granite and limestone (all the rest))
E = Pompei, Italy [27]	**6** = *Vestibulum, Tepidarium and Swimming pool* of the Herculaneum Suburban Baths (green area, sampled from plaster (Vestibulum), marble (Tepidarium) and mortar (swimming pool))
F = Rome, Italy [28,31]	**7** = *Trevi Fountain* (urban area, marble and travertine); the presented genera in the tables were observed on the same type of stone samples immersed in the fountain basin
	8 = *Fuga's Fountain* from the Botanical Garden of Rome (green area)
G = Gujarat, India [32]	**9** = *Fountain Reservoir of Seth Sarabhai's Garden* (green area)
	10 = *Planet of Peace Fountain* (urban area)
	11 = *Hviezdoslav's Fountain* (urban area)
	12 = *Ganymede's Fountain* (urban area)
	13 = *Friendship Fountain* (urban area)
	14 = *Veil Fountain* (urban area)
I = Granada, Spain [21,23–26]	**15** = *Patio Sultana* (Alhambra Complex) (green area, made of Sierra Elvira stone)
	16 = *Lindaraja Fountain* (Alhambra Complex) (green area, made of marble (pedestal) and Sierra Elvira stone (basin))
	17 = *Patio Naranjos* (Alhambra Complex) (green area)
	18 = *Patio Reja* (Alhambra Complex) (green area)
	19 = *Lions Fountain* (Alhambra Complex) (green area)
	20 = *Mexuar Fountain* (Alhambra Complex) (green area)
	21 = *North "guitar" Fountain* of the Court of the Myrtles (Alhambra Complex) (green area)
	22 = *South "guitar" fountain* of the Court of the Myrtles (Alhambra Complex) (green area)
	23 = *East channel of the Court of the Myrtles* (Alhambra Complex) (green area)
	24 = *"Guitar" Fountain of the Ladies Tower* (Alhambra Complex) (green area)
	25 = *Jardín de los Adarves east basin* (Alhambra Complex) (green area)
	26 = Mirador del Partal basin (Alhambra Complex) (green area)
	27 = *Glorieta del Secano Fountain* (Alhambra Complex) (green area)
	28 = *Jardines Bajos South Fountain* (Alhambra Complex) (green area)
	29 = *Escalera del Agua handrail* (Alhambra Complex) (green area)
	30 = *Fuente del Tomate* (Alhambra Complex) (green area)
	31 = *Ángel Ganivet pool Fountain* (Alhambra Complex) (green area)
	32 = *Carlos V basin* (Alhambra Complex) (green area)
	33 = *Washington Irving basin* (Alhambra Complex) (green area)
	34 = *Puerta de las Granadas basin* (Alhambra Complex) (green area)
	35 = *Right fountain of the 2nd terrace of the Carmen de Bellavista* (Alhambra Complex) (green area)
	36 = *Bibatauín Fountain* (urban area)
J = Seville, Spain [25]	**37** = *Reales Alcázares Fountains* (unspecified fountains) (green area)
K = Washington, DC, USA [22]	**38** = *The outside pool of the National Museum of African American History and Culture*, Smithsonian Institution
	39 = *The pool in the plaza of the Hirshhorn Museum*, Smithsonian Institution
	40 = *Large Acanthus Fountain*, Smithsonian Institution
	41 = *Keith Fountain*, Smithsonian Institution
	42 = *The pool in front of the National Museum of American History*, Smithsonian Institution
	43 = *The fountain on the northwest corner of the National Museum of the American Indian*, Smithsonian Institution
	44 = *The fountain outside the National Gallery*, which flows into the cafeteria
	45 = *The fountain located in front of the west façade of the West Building of the National Gallery*
	46 = *Girl Water Lilies Fountain of the National Gallery*

Table 2. Cyanobacteria observed in different artistic fountains (see Table 2 for details of the location), where CO = Columbia, IT = Italy, ES = Spain and USA = United States.

Location	Cyanobacteria in the Worldwide Fountains																																	
Country	CO			IT										ES																USA				
City (A–K)	A	B	C		D	E		F									I									J				K				
Fountain (1–46) Microorganisms	1	2	3	4	5	6	7	8	15	16	18	19	21	22	23	24	26	27	28	29	30	31	32	34	35	36	37	38	39	40	41	43	44	46
Aphanocapsa			x	x					x																									
Aphanothece				x						x																								
Borzia				x																														
Calothrix		x	x	x	x				x		x			x											x		x			x	x		x	
Chamaesiphon					x					x	x		x				x	x			x		x			x								
Chlorogloea														x						x	x					x								
Chroococcidiopsis						x			x																	x				x	x			
Chroococcopsis					x					x														x										
Chroococcus			x	x	x		x	x	x																		x							
Cyanosarcina												x						x																
Dichothrix													x																					
Gloeobacter				x																														
Gloeocapsa			x	x	x	x		x	x																	x								
Gloeocapsopsis						x																												
Gloeothece					x	x																												
Leptolyngbya			x	x		x			x	x	x	x	x	x	x	x				x		x						x	x	x	x	x		x
Lyngbya					x		x	x	x																									
Microcoleus					x																													
Microcystis	x						x																											
Myxosarcina					x															x												x		
Nostoc	x			x				x	x																									
Oscillatoria	x	x			x	x	x																											
Phormidium		x				x	x	x	x	x		x	x	x		x		x	x		x				x	x	x							
Plectonema					x																													
Pleurocapsa					x				x				x	x	x											x								
Pseudanabaena						x			x										x												x			
Pseudophormidium			x	x		x			x																									
Rivularia				x																														
Schizothrix				x																														
Scytonema						x																												
Stanieria									x																									
Symploca											x					x											x							
Synechococcus					x																													
Synechocystis					x																													

Table 3. Green algae observed in different artistic fountains (see Table 3 for details of the location), where CO = Columbia, IT = Italy, IN = India, SK = Slovakia, ES = Spain and USA = United States.

Location	Green Algae in the Worldwide Fountains																																											
Country	CO				IT				IN			SK												ES																	USA			
City (A–K)	A	B	C		D	E	F		G			H											I											J							K			
Fountain (1–46) Microorganisms	1	2	3	4	5	6	7	8	9	10	11	12	13	14	15	16	17	18	19	20	21	22	23	24	25	26	27	29	30	31	32	33	34	36	37	38	39	40	41	42	43	44	45	46
Apatococcus						x									x	x						x													x	x			x		x			x
Bracteacoccus															x					x					x	x	x		x	x	x	x	x		x									
Chlamydomonas						x				x	x	x	x	x																											x			
Chlorella			x	x	x	x	x	x		x	x	x	x	x	x				x					x						x					x									
Chlorococcum					x	x										x			x																			x	x				x	
Chlorosarcina											x							x										x							x									
Chlorosarcinopsis						x				x	x	x	x	x		x			x	x	x															x			x					
Choricystis										x	x	x	x	x											x																			
Coccomyxa						x					x	x	x	x																														
Cosmarium			x						x	x	x	x	x	x	x																			x										
Dilabifilum															x																													
Euglena	x																																											
Gloeocystis																							x																					
Gongrosira															x																													
Haematococcus											x		x	x																														
Klebsormidium														x		x	x		x																									
Leptosira																												x																
Microthamnion						x																																						
Monoraphidium				x						x		x	x																															
Muriella						x							x	x																														
Oocystella										x	x	x	x	x																														
Oocystis																																					x					x		
Palmella			x																																									
Pediastrum									x	x	x	x	x	x																														
Planophila																																		x	x									
Pleurastrum															x									x																				
Pseudochlorella						x					x			x																														
Pseudopleurococcus																												x																
Selenastrum	x																																										x	
Scenedesmus							x		x	x	x	x	x	x				x					x	x														x	x	x	x	x	x	
Spirogyra	x	x			x									x																														
Stichococcus						x	x	x											x									x						x	x		x			x	x			
Tetracystis																																			x									
Trebouxia					x							x																																
Ulothrix					x				x																									x	x						x			

Table 4. Diatoms observed in different artistic fountains (see Table 1 for details of the location), where CO = Columbia, IT = Italy, IN = India, ES = Spain and USA = United States.

Location	**Diatoms in the Worldwide Fountains**																					
Country	**CO**	**IT**						**IN**	**ES**										**USA**		
City (A–K)	**A**	B	C		D	F		**G**					**I**						J	**K**	
Fountain (1–46) **Microorganisms**	**1**	2	3	4	5	7	8	**9**	15	16	18	19	20	21	22	23	26	28	36	37	**38**	**43**
Achnanthes	x		x					x	x			x			x	x			x			
Amphora						x		x														
Cymbella	x							x	x						x					x		
Diatoma	x								x													
Epithemia			x					x												x		
Frustulia	x					x																
Gomphonema	x	x						x														
Melosira								x												x		
Navicula	x	x	x	x				x	x	x	x	x	x	x	x		x	x	x	x	x	x
Nitzschia	x		x			x		x	x		x		x						x	x		x
Pinnularia	x					x		x		x												
Surirella		x						x														
Synedra	x					x		x		x										x		
Tabellaria	x																		x			

The collection of the phototrophic biomass is usually made by using sterile scalpels and tubes of different sizes. When possible, in the case of chasmolithic or endolithic developments, small pieces of stone substrata containing microorganisms can be sampled as well. A fixation with formalin (2%, final concentration) is recommended in order to preserve the morphology of the phototrophs [23]. The Lugol's iodine solution (1% v/v) is also used for preventing contamination [5], being also a useful test under the microscope, as it gives a dark purple/brown/blue black stain to the cells containing starch (green algae and diatoms), favoring the sedimentation of the cells and their easier observation as well [33].

The microscopic observations of the samples can be made in transmission light, and good quality images can be obtained by using DIC (differential interference contrast) microscopy. The fresh samples will often contain more than one genus. The sample preparation for the microscopic analysis can be summarized as following: (i) first of all, place a drop of water on the center of a clean slide; (ii) take a very small part of the sample (e.g., with the help of a needle or tweezers) and place it inside the drop of water; if the sample is solid or too thick (e.g., a dense mat or a crust) it must be crushed and scattered in the water drop in order to be able distinguish individual cells by the light passing through the sample; it is important to not mechanically break the clusters before the slide preparation, as confusion may occur with the non-colonial genera; (iii) once the sample is uniformly distributed, place slowly the cover on top by firstly putting in contact its outer edge with the water drop. If the water overflows, gently remove the excess with a paper towel. When diatoms are present, a cleaning procedure of the frustules can be used before preparing the slides [34,35]. This procedure involves a careful use of some acids (nitric acid, hydrogen peroxide), able to remove the organic matter and carbonates from the sample, and therefore to highlight the morphological characteristics of the diatoms. Once prepared, the slides can be observed under different magnification levels (10×, 20×, 40×, 50× or 100×) in order to distinguish many of the morphological features of the cells. Some of the main features that would be observed are related to the shape of the cells, the presence or absence of the sheaths, the division type, the end cells morphology in the case of the filamentous cyanobacteria, the presence of motility phenomena, the plastids position and their shape in the case of green algae. It is possible to make deeper investigations by evaluating the ratio within the main phototrophic groups, as Popović et al. [36] describe a detailed procedure for the mixed phototrophic biofilms.

2.2. Identification Keys

The keys derive from the specific literature survey on the taxonomy of cyanobacteria [11–19], green algae [11,13] and diatoms [11,13,21] and from the authors' collected material. The keys are dichotomous. Each number describes a couple of contrasting and distinctive characteristics. The reader makes a choice between the two options and follows the indications for the next couplet and so on. The dichotomous chain finishes with finding the genus. It should be specified that many species of cyanobacteria exhibit variable morphology, and their identification till the species level may need specific ultrastructural and molecular features, a fact that it was not considered necessary for the aim of this work and therefore they were not used in this key.

Five plates containing the illustrations of the main common genera were realized in watercolor technique (Arches cotton paper 300 g/m^3, Sennelier and Mijello water colors) by F. Bolívar-Galiano, according mainly to our microscopic observations.

3. Results

3.1. First Screening

The first screening can be undertaken by naked eye or by using a portable microscope with about 20× magnification which can help to choose the sampling area and to make a rough morphological discrimination, through the inspection of the color and morphological aspects of the biological colonization (mat aspect with visible filaments, patina aspect with uniform color, spots like growth, etc.). A good camera with a macro objective can also be very useful.

In this step, it is difficult to assign the sample to a single phototrophic group, as on monumental fountains and other wet stone surfaces there are developing communities of microorganisms, which are a mix of different cyanobacteria and/or green algae and/or diatoms, as can be seen in some fresh samples observed under the microscope (Supplementary Materials S1–S5). Diatoms prefer high levels of humidity and are more often associated with green algae, and less only with cyanobacteria, but they can be found beside algae and cyanobacteria as well. Often, the color gives an idea about the possible phototrophic presence, but it is never conclusive [37].

1a visible vivid green, light green, green or brownish phototrophic growth........................2
1b visible dark green, deep brown, deep grey or black phototrophic growth.....................3
2a with a mucilaginous, wet or powdering aspect, it seems not well adhered to the stone surface..4
2b with an aspect of patina, pellicle or mat, that seems adhered to the stone surface..........5
3a dark green, dark grey or dark brown wet formations, with filamentous or spherical aspect...................................see *3.2. Cyanobacteria (blue-green algae)* and *3.3. Green Algae*
3b dark patina with intergranular growth on dry surfaces..................see *3.2. Cyanobacteria*
4a vivid and light green color..see *3.3. Green Algae*
4b green, yellow-green or light brown color, sometimes with aspect of pustule...see *3.3. Green Algae* and *3.4. Diatoms*
5a pellicles more or less gelatinous......................see *3.2. Cyanobacteria* and *3.3. Green Algae*
5b aspect of mats and patina, with different thicknesses, dark green or dark blue-green color with many filaments...see mainly *3.2. Cyanobacteria*

3.2. Cyanobacteria (Blue-Green Algae)

They are prokaryotic phototrophic microorganisms, without chloroplasts (plastids), with an immense morphological variability, from free and isolated cells to very hard and consistent assemblages. Moreover, the Lugol solution used for sample preservation, does not stain the cells blue/black. Cyanobacteria have a characteristic blue-green color, hence their common name of blue-green algae. They can be divided in two main groups: I. simple cyanobacteria and II. filamentous cyanobacteria.

I. simple cyanobacteria (unicellular, mucilaginous groups, strong adherent group of cells) (Figures 1 and S6 and [23,38])

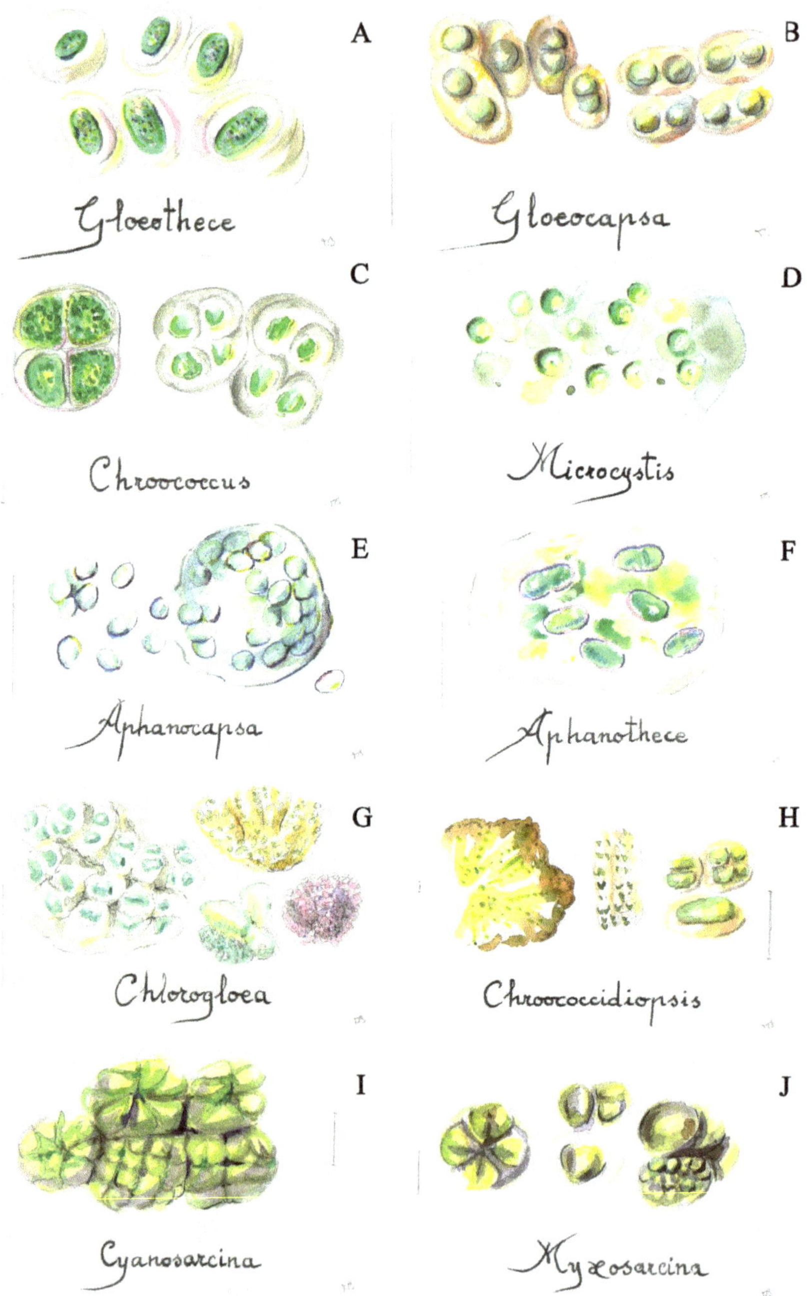

Figure 1. Some simple cyanobacteria (blue-green algae): (**A**) *Gloeothece* (4a), (**B**) *Gloeocapsa* (5a), (**C**) *Chroococcus* (5b), (**D**) *Microcystis* (7a), (**E**) *Aphanocapsa* (7b), (**F**) *Aphanothece* (8a), (**G**) *Chlorogloea* (9a), (**H**) *Chroococcidiopsis* (9b), (**I**) *Cyanosarcina* (12a), (**J**) *Myxosarcina* (12b). Scale bar 10 μm. The number in the brackets corresponds to the one indicated in the identification key.

1a mucilaginous or compact groups of cells............2
1b cells lacking mucilage............10
2a irregular agglomerations of cells embedded in mucilage............3
2b agglomerations of cells with a regular packet-like arrangement............13
3a concentrically layered gelatinous sheaths around the cells (with distinct or indistinct lamellation)............4
3b sheaths without clearly concentric lamellation............6
4a ovoid to rod shaped cells, well delimitated and lamelled sheaths............*Gloeothece*
4b spherical to hemispherical cell shape, with colorless or light yellow-brown sheaths....5
5a sheaths wide and vesiculous............*Gloeocapsa*
5b sheaths thin and usually colorless, usually small colonies (2–16 cells, with the shape of a half or a quarter of sphere after division)............*Chroococcus*
6a irregular colonies forming amorphous masses............7
6b irregular colonies with their cells distributed in the mucilage............8
7a spherical cells............*Microcystis*
7b spherical or slightly elongated cells (oval, ovoid)............*Aphanocapsa*
8a ellipsoidal cells with homogeneous sheaths............*Aphanothece*
8b spherical cells............9
9a cells with an irregular-rounded to polygonal-rounded outline, and with the margin of the colony in a more or less radial row............*Chlorogloea*
9b cells more or less uniform in size in the colony............*Chroococcidiopsis*
10a cells ovate to oblong, single or in small colonies............11
10b spherical and solitary cells............*Synechocystis*
11a cells with a dimension of (1.5)3–15(40) $\times$ 0.4–3(6) μm............*Synechococcus*
11b very small cells, up to 1.5 $\times$ 3 μm............*Gloeobacter*
12a agglomerations of cells with a more regular arrangement (cubical)............*Cyanosarcina*
12b agglomerations of cells with a less regular arrangement............*Myxosarcina*

II. Filamentous cyanobacteria (linear groups of cells called trichomes, with one or more rows, surrounded or not by sheaths) (Figures 2, S7 and S8 and [23,38])

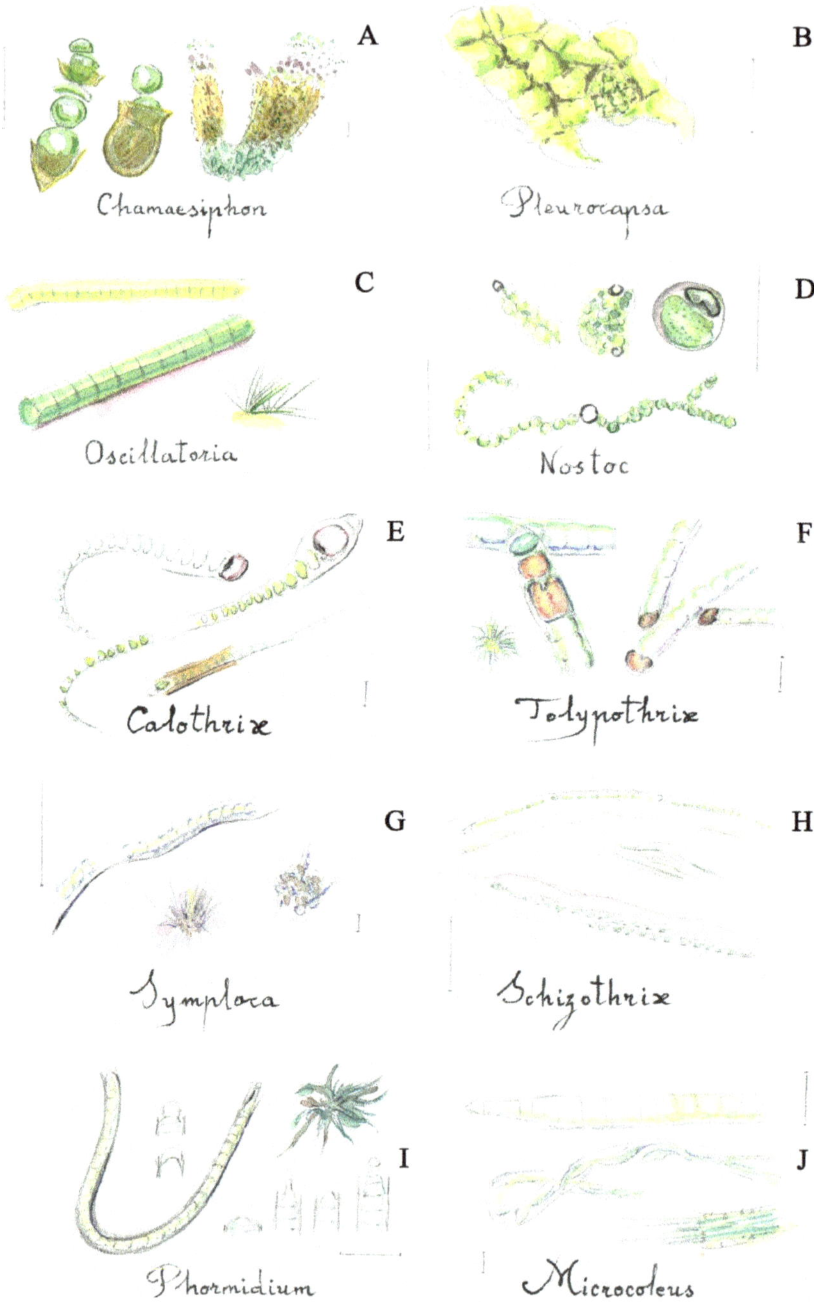

Figure 2. Some filamentous cyanobacteria (blue-green algae): (**A**) *Chamaesiphon* (2a), (**B**) *Pleurocapsa* (2b), (**C**) *Oscillatoria* (5a), (**D**) *Nostoc* (6a), (**E**) *Calothrix* (9a), (**F**) *Tolypothrix* (10b), (**G**) *Symploca* (13b), (**H**) *Schizothrix* (14a), (**I**) *Phormidium* (16a), (**J**) *Microcoleus* (17a). Scale bar 10 μm. The number in the brackets corresponds to the one indicated in the identification key.

1a trichomes not clearly developed, compact groups of cells....2
1b trichomes clearly developed, isolated or forming mats or pellicles....3
2a short apical row of cells easily damaged, basal cells with colored sheath, forming macroscopic brown spots....*Chamaesiphon*
2b irregular groups of rows—pseudofilaments—built by polygonal cells....*Pleurocapsa*
3a trichomes without sheath or with a very diffluent sheath....4
3b trichomes with a visible sheath....6
4a trichomes very short, up to 8 cells (rarely 16)....*Borzia*
4b trichomes with a higher number of cells 5
5a trichomes with very short cells, coin shaped and close together, visible moving filaments under the microscope....*Oscillatoria*
5b trichomes with long cells more distant from one to another....*Pseudanabaena*
6a trichomes forming firm colonies with mucilage ....*Nostoc*
6b trichomes forming compact clusters with a covering character 7
7a presence of differentiated cells (heterocytes) in addition to the rest of the cells 8
7b all cells with the same morphology, sometimes apical cells are differentiated....11
8a heterocytes present at the base of polar filaments, filaments narrow into the shape of a hair....9
8b heterocytes intercalated along the filaments, filaments not narrow....10
9a filaments with thin sheaths, no branching....*Calothrix*
9b filaments with wide sheaths, false-branching ....*Dichothrix*
10a filament usually with rounded cells, no branching....*Nostoc*
10b filaments with rectangular or squared cells and false branching, usually prostrated....*Tolypothrix*
11a one trichome within each sheath 12
11b several trichomes within each sheath 17
12a consistent sheaths with apical cells not differentiated, filaments can form mats....13
12b mucilaginous sheaths with apical cells usually differentiated....15
13a mats formed exclusively by prostrated filaments....14
13b mats formed by a base of prostrated filaments that form peripheral erect bundles with a conical shape, cells lack the calypra....*Symploca*
14a strongly intertwined filaments with false branching and wide sheaths, and sometimes more than one trichoma....*Schizothrix*
14b very intertwined filaments forming less consistent mats, usually between mineral concretions....*Lyngbya*
15a squared or slightly rectangular cells....16
15b cylindrical cells longer than wide, less than 3 μm in width, unbranched or with false branching....*Leptolyngbya*
16a filaments with no branching, sheath sometimes difficult to see....*Phormidium*
16b filaments with false branching....*Pseudophormidium*
17a numerous trichomes within each sheath....*Microcoleus*
17b one or few trichomes within each sheath....18
18a few-celled and short trichomes, constricted at cross walls, with a more or less cylindrical shape....*Pseudanabaena*
18b trichomes in vegetative state always within distinct sheaths, filaments with typical "scytonematoid" (double and single) false branching, forming slimy mats....*Plectonema*

3.3. Green Algae (Chlorophyta)

This division contains very different organisms regarding the morphological, dimensional and reproductive aspects. They have a typical vivid light green color, but the cells with starch will stain dark purple/brown/blue black in Lugol's iodine solution (if used for preserving the sample). They contain chloroplasts (plastids) and can be divided into two main groups: I. simple green algae (and green flagellate) and II. filamentous green algae.

I. Simple green algae (and green flagellate Euglena) (Figures 3 and S9 and [21,23,38])

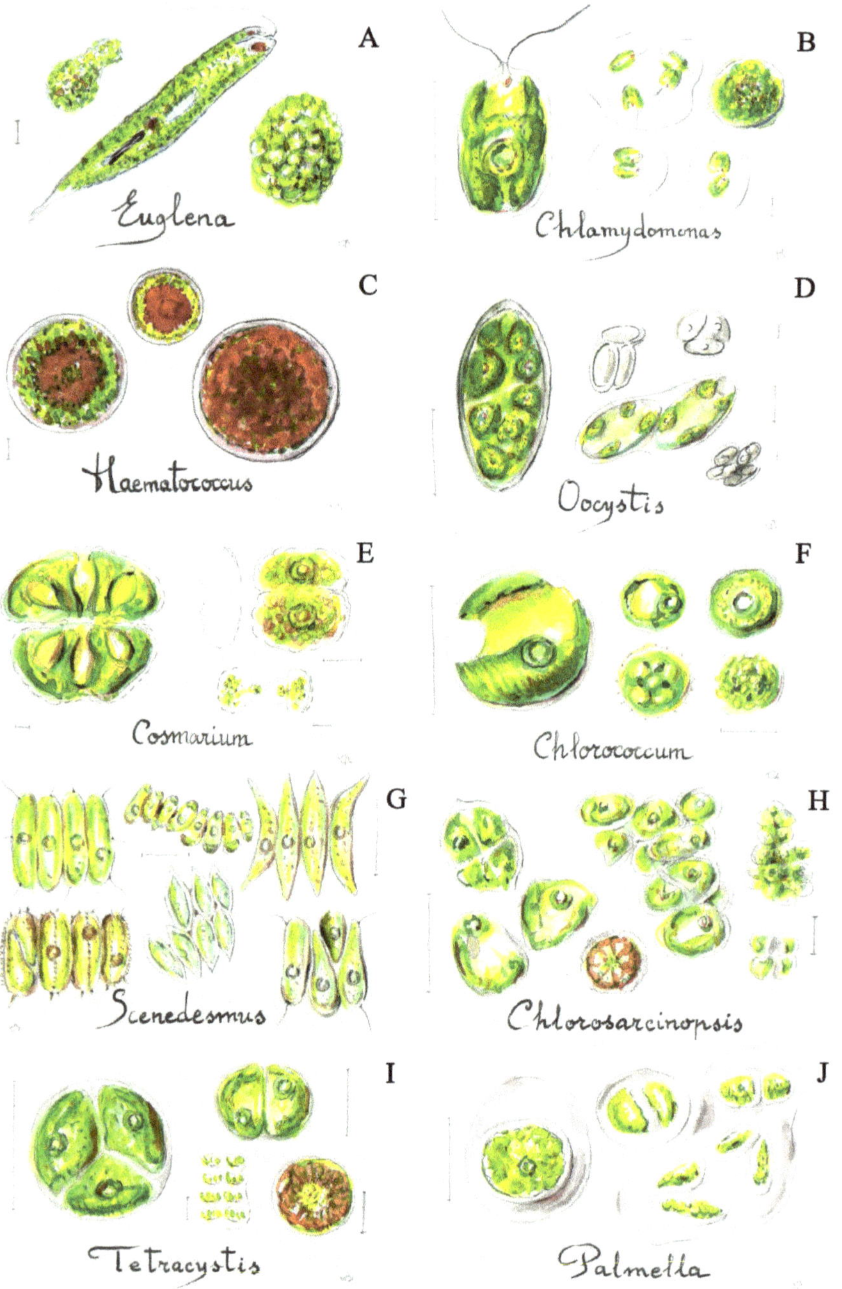

Figure 3. Some simple green algae: (**A**) *Euglena* (3b), (**B**) *Chlamydomonas* (4a), (**C**) *Haematococcus* (4b), (**D**) *Oocystis* (8a), (**E**) *Cosmarium* (12a), (**F**) *Chlorococcum* (13b), (**G**) *Scenedesmus* (17b), (**H**) *Chlorosarcinopsis* (19a), (**I**) *Tetracystis* (19b), (**J**) *Palmella* (20a). Scale bar 10 μm. The number in the brackets corresponds to the one indicated in the identification key.

1a solitary cells....2
1b cells forming colonies....14
2a motile cells 3
2b non-motile cells....5
3a cells with two flagella and with a rigid cellular surface....4
3b cells with flagella but with elastic (deformable) cellular surface....*Euglena*
4a solitary cells with an oval or globular shape....*Chlamydomonas*
4b cells with a more or less globular shape, with a red spot or totally red-colored in adverse conditions....*Haematococcus*
5a cells not closely grouped, more or less spherical or globular....6
5b cells closely and continuously grouped....14
6a elliptical, egg-shaped cells....7
6b spherical cells, isolated or arranged in unstructured groups....9
7a cells clearly elongated, fusiform....*Monoraphidium*
7b cells are ellipsoidal or globular....8
8a cells clearly elliptic, sometimes lemon-shaped, colonies delimitated by a parental wall....*Oocystis*
8b ovoid cells, almost globular, solitary or in irregular mucilaginous colonies...*Coccomyxa*
9a many parietal plastids inside the cell, without pyrenoids....10
9b few plastids inside the cell, usually with pyrenoids....11
10a solitary multinucleated large cells, maximum 85 μm in diameter....*Bracteococcus*
10b spherical cells with a diameter of 5–18 μm....*Muriella*
11a spherical cells with a dimension of 2–25 μm, with a huge star-shaped plastid, usually as a lichen symbiont....*Trebouxia*
11b ellipsoidal, globular or sub-spherical cells....12
12a cells with a median constriction dividing them into two semi-cells....*Cosmarium*
12b cells without constriction....13
13a cells with a cup- or plated-like plastid, maximum diameter of 20 μm....*Chlorella*
13b spherical, sometimes ellipsoidal cells with a variable diameter inside the same population (from 10–15 to 100 μm), with the plastid filling the cell....*Chlorococcum*
14a groups of cells that are close together....15
14b groups of usually separated cells forming mucilaginous structures....20
15a small and bidimensional groups of cells....16
15b large groups of cells, not bidimensional 18
16a cells are arranged on a single layer, with a two-dimensional structure 17
16b cells are arranged in groups of four (sometimes in a clear gelatinous matrix) or isolated, with strip- or plated-like plastids....*Pseudochlorella*
17a cells are globular ....*Planophila*
17b cells are elliptical, spindle-shaped or pear-shaped, forming groups of 4–8 cells (or multiples of four), arranged linearly or in zig-zag ....*Scenedesmus*
18a cubical shape of the colony....*Chlorosarcina*
18b non-cubic colonies....19
19a cells are very compacted, forming angular-shaped colonies....*Chlorosarcinopsis*
19b globular shape of the colony, with its cells isolated or in small groups (two or four cells) usually tetrahedron-like shaped....*Tetracystis*
20a globular or amorphous colonies, cells with a diameter of 10–15 μm....*Palmella*
20b globular, tetrahedron-like or irregular shape of the colony, with a consistent gelatinous matrix and with every cell enclosed by a visible membrane, the diameter varies between 3 and 23 μm ....*Gloeocystis*

II. Filamentous green algae (Figures 4 and S9 and [21,23,38])

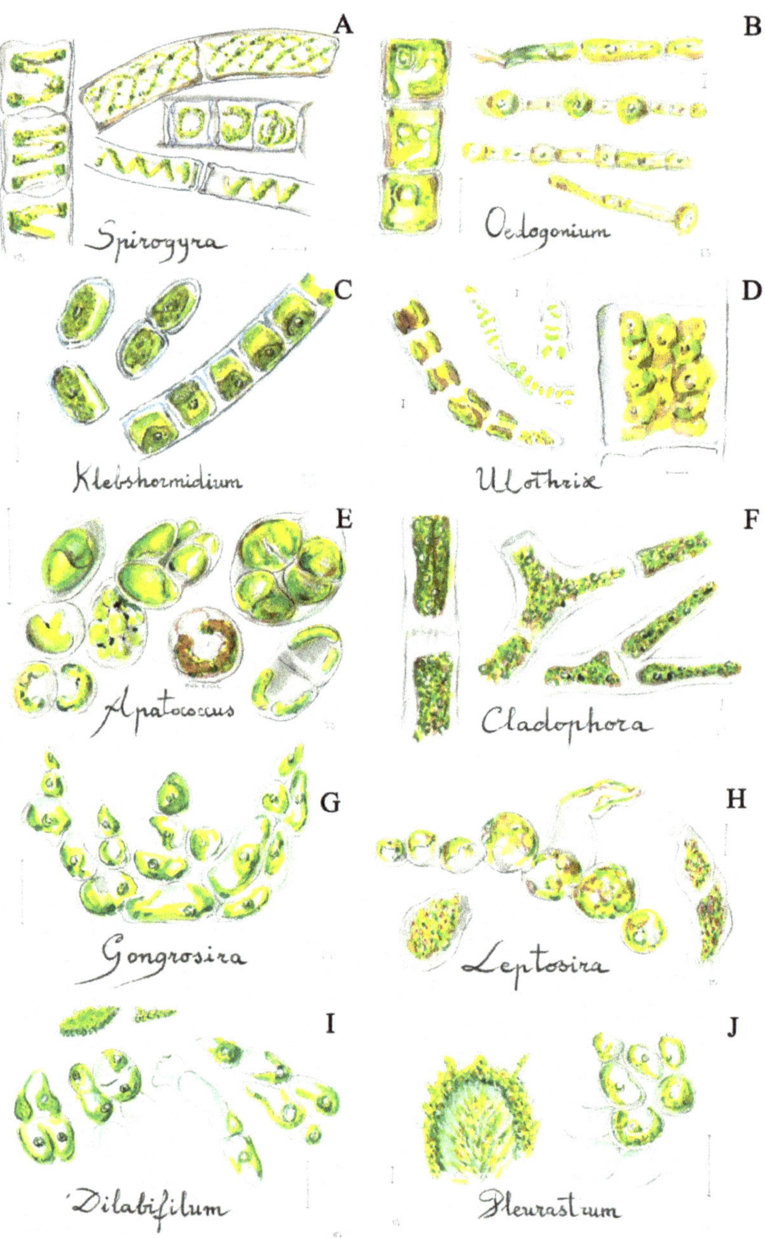

Figure 4. Some filamentous green algae from the C2 group: (**A**) *Spirogyra* (2a), (**B**) *Oedogonium* (3a), (**C**) *Klebsormidium* (4b), (**D**) *Ulothrix* (4a), (**E**) *Apatococcus* (5a), (**F**) *Cladophora* (6a), (**G**) *Gongrosira* (7a), (**H**) *Leptosira* (8a), (**I**) *Dilabifilum* (9b), (**J**) *Pleurastrum* (9b). Scale bar 10 μm. The number in the brackets corresponds to the one indicated in the identification key.

1a filaments not branched..2
1b filaments branched, sometimes forming very compact structures 5
2a plastids with the shape of a spiral...*Spirogyra*
2b plastids with another shape 3
3a presence of polar rings and intercalary widened cells (oogonia) ....*Oedogonium*
3b all cells with similar morphology and without polar rings....................................4
4a long filaments fixed at their base, with a stable structure, annular plastids.....*Ulothrix*
4b filaments that disarticulate easily, usually short, annular or elliptic plastids, usually aerophytic..*Klebshormidium*
5a short filaments forming packet-like colonies, sometimes cubical in the first stages, cells globular or elongated..*Apatococcus*
5b filaments clearly uniseriate, linear or branched..6
6a long filaments made of long cylindrical cells with laminated wall and reticulate plastids with numerous pyrenoids..*Cladophora*
6b short filaments or small groups of cells usually with a prostrate base and a small erect part, parietal plastids with few or no pyrenoids..7
7a cylindrical or club-shaped cells with a blunt apex, usually tangled filaments, built crustose lime impregnated ....*Gongrosira*
7b globular or cylindrical cells forming compact groups, sometimes without a clear filament structure..8
8a globular cells, parietal plastids without pyrenoids, short filaments without branches or with short branches of 1–2 cells..*Leptosira*
8b cylindrical or globular cells, with pyrenoids, from irregular groups to small filaments.9
9a cylindrical-elongated cells with short branches ....*Dilabifilum*
9b cylindrical or subglobular cells, forming both irregular and dense groups and branched filaments ....*Pleurastrum*

3.4. Diatoms (Bacillariophyta)

They are unicellular reddish brown or golden algae (Figures 5 and S10 and [23,38]). The specific characteristic of this group is the presence of an external layer (the cell wall) made of silica that is called frustule. They are common on permanent wetted surfaces, and on monumental fountains the most encountered diatoms are those with bisymmetrical or monosymmetrical valves, one overlapping the other. The diatoms with radial symmetry are less common.

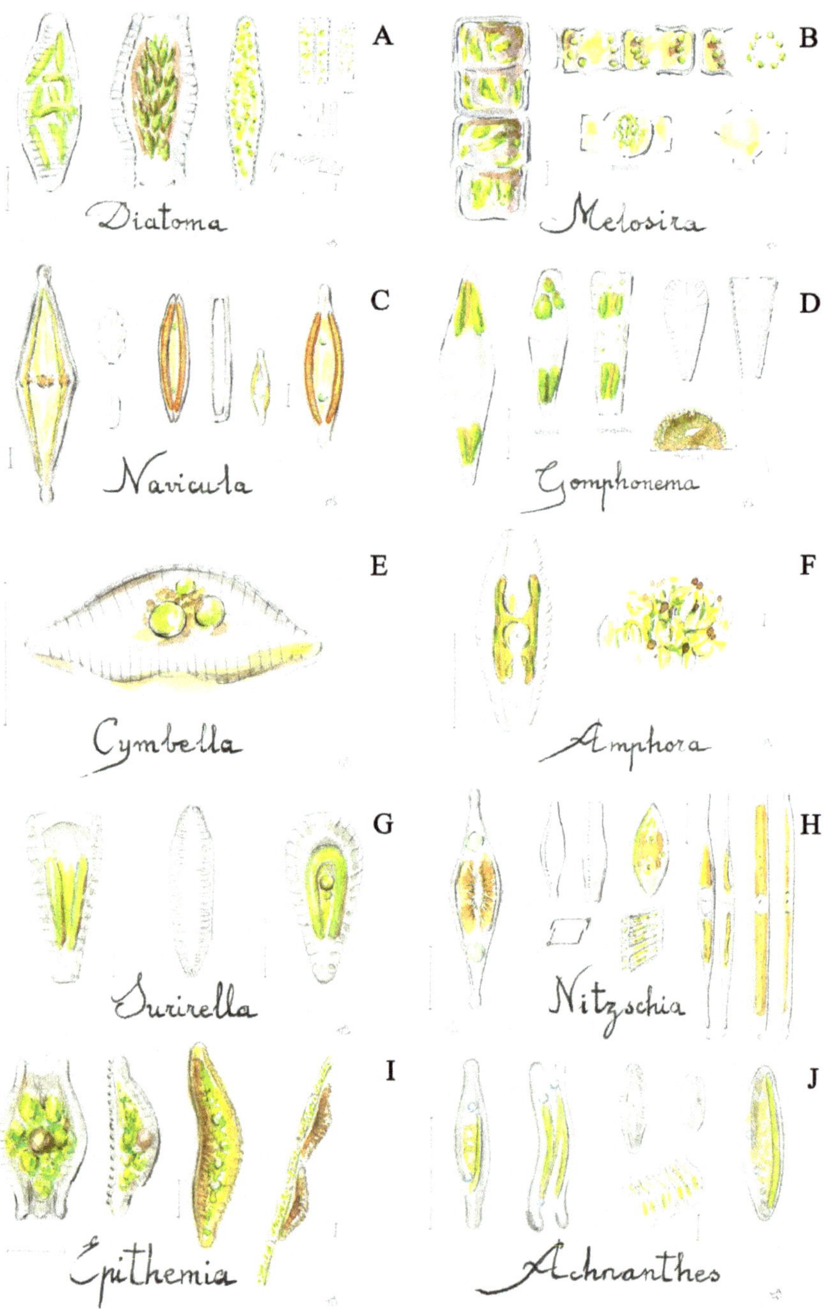

Figure 5. Some diatoms: (**A**) *Diatoma* (2a), (**B**) *Melosira* (2b), (**C**) *Navicula* (8a), (**D**) *Gomphonema* (9b), (**E**) *Cymbella* (10a), (**F**) *Amphora* (10b), (**G**) *Surirella* (11b), (**H**) *Nitzschia* (12a), (**I**) *Epithemia* (12b), (**J**) *Achnanthes* (13a). Scale bar 10 μm. The number in the brackets corresponds to the one indicated in the identification key.

1a cells with a colonial organization, united in filament-like structures, numerous discoidal plastids........2

1b solitary cells, sometimes forming colonies, with valves of different shapes (elliptic, lanceolate, oblong, etc.), but never circular and with no radial symmetry, two lateral plastids........3

2a short cells, the filaments are easily separated in small groups of cells, valves symmetric both on the frontal and lateral faces, septa are lacking in the frustules........*Diatoma*

2b long cells, valves with a circular shape (only visible in valvar view of isolated cells), the cells form filaments often found inside the shady fountains........*Melosira*

3a both valves with a raphe (structure composed by two slits or fissures), usually with a striation from the raphe to marginal part........4

3b both valves without a raphe........14

4a the raphe is located in the middle of both valves........5

4b the raphe is located in a different part........12

5a the raphe is located on both valves........6

5b the raphe is located along the margin of the valves........11

6a bi-symmetric valves, central raphe........7

6b mono-symmetric valve, central or lateral raphe........9

7a striation narrow or not very wide that is not visible........8

7b very wide striations, the raphe bent at both ends........*Pinnularia*

8a thin striation, visible in dead cells, valve elliptical to boat-shaped........*Navicula*

8b very thin striations (horizontal and vertical), difficult to see even in dead cells....*Frustulia*

9a the valves are longitudinally asymmetric, raphe marginal........10

9b the valves are transversely asymmetric, with the upper part broader than the lower one, with a trapezoidal girdle view........*Gomphonema*

10a the valve margins are different, one convex and the other straight or slightly curved, raphe parallel to the straight margin and striations do not strongly radiate in the center ....*Cymbella*

10b elliptic or oblong valves vision, intercalary bands and striations are imperceptible........*Amphora*

11b the raphe is present on two sides of each valve, with thick striation and a big central channel........*Surirella*

11a the raphe is present on one side of each valve........13

12a hyaline valves, usually linear with lateral refractory dots (fibulae)........*Nitzschia*

12b the raphe forms a peak in the ventral part of the valve........*Epithemia*

13a linear-elliptic valve, with a bent girdle view........*Achnantes*

13b elliptic and almost circular valve, straight girdle view........*Cocconeis*

14a needle-shaped valves with parallel and thin striations, lacking septa (a silica internal sheet, occluding a portion of the frustula)........*Synedra* (*Ulnaria*)

14b the frustules present septa, observed in girdle view, forming filaments........*Tabellaria*

4. Discussion

4.1. Fountains' Phototrophic Biodiversity

Fountains are ideal environments for the rapid development of microorganisms, which induce the occurrence of biodeterioration processes and algal blooms that should be controlled in the case of artistic structures. Besides the bioreceptivity of the material and the characteristics of the environmental factors, the water origin is another element that highly influences the biodiversity of the colonizers on monumental fountains. The water source used for their alimentation can be driven from a public water line (e.g., mainly in the city centers) or can be directly supplied from a nearby river (e.g., in some gardens) [4,7,8,21,23,24,27]. The latter type of water source induces a greater biodeterioration risk for the monumental fountains.

A review of the main phototrophic microorganisms detected by various authors on this kind of artefact (Table 1) is presented in the Tables 2–4. Table 1 contains the legend of Tables 1–3. The microorganisms are divided in the following three groups: cyanobacteria (Table 2), green algae (Table 3) and diatoms (Table 4) and most of them are included in the identification keys. Very few studies have been undertaken on the molecular identification of the phototrophs dwelling on cultural fountain structures [23,27]. The identification to the species level needs axenic cultures of the isolated organisms, increasing the complexity of these types of analyses. Moreover, many of these phototrophs (mainly cyanobacteria) have a slow growth and their isolation is very time consuming. Thanks to the recent developments in the molecular biology domain, new techniques such as NGS (next-generation sequencing, called also high throughput sequencing—HTS) and metagenomics approaches give now valid and cost-effective solutions in the cultural heritage field [39,40]. The latter technique, called also community genomics, contributes with new insights for a better understanding of both the role and the structure of the microbial communities colonizing stone artefacts. However, these studies [23,27] confirmed the presence of various cosmopolitan phototrophic microorganisms with wide tolerance ranges concerning, for example, pH and temperatures. Many of these common genera (such as *Calothrix*, *Phormidium*, *Oscillatoria*, *Cosmarium*, *Scenedesmus*, *Ulothrix*, *Navicula*, *Nitzchia*, etc.) can be easily identified by their morphological characteristics by using fresh samples, a transmission microscope and an easy to use identification key. It should be noted that some genera can be easily confounded in between [41,42] and therefore it is important to pay attention to the minimal morphological differences (e.g., *Gloeocapsa* with *Aphanocapsa*, when the *Gloeocapsa* individual sheath cannot be distinguished in a colony; *Aphanocapsa* with *Microcystis*, as *Aphanocapsa* can be round-shaped as well; *Cyanosarcina* with *Myxossarcina* as the shapes of the single colonies in the agglomeration of cells are not always very clear; some small sized diatoms can be confused with *Navicula* or *Nitzschia*).

The phototrophic microorganisms form different types of biocenosis (biotic communities) and it can be stated that, almost always, the phototrophic growth will be a mix of different genera as can be usually observed when analyzing fresh samples (Figures S1–S5 Supplementary Materials), containing even other types of organisms such as bacteria or fungi. The stone support induces a certain bioreceptivity for the phototrophic colonization. Stone materials with higher porosity (travertine—fountain no. 7; sandstones—fountain no.3; cementitious materials—fountains no.4 and no.6; volcanic rocks—fountain no.1, see Table 1) are easily colonized by the phototrophic organisms with respect to the more compact stone materials (granite—fountain no. 5; marble—fountains no.5, no. 6, no.7 and no.15). The limestones and sandstones are more sensitive to biodeterioration with respect to granitic stones [43]. The common genera observed in fountains worldwide can have an epilithic growth forming mucous (*Apatococcus*, *Phormidium*) or fibrous mats (*Cladophora*, *Gongrosira*, *Melosira*, *Pleurastrum*). The existing fissures and cracks are colonized by chasmo-endoliths (mainly simple cyanobacteria and green algae) while the internal pores are colonized by cryptoendoliths (*Chroococcidiopsis*). Some genera (*Chroococcidiopsis*, *Chroococcus*, *Gloeocapsa*, *Klebsormidium*, *Schizothrix*, *Synechococcus*) can have an endolithic growth, actively penetrating inside the material, being therefore more harmful for the stone support [44,45].

Green algae and cyanobacteria present a greater diversity compared to diatoms when the number of most frequent genera in the three groups is analyzed (Tables 2–4). On the studied fountains, *Navicula* spp. and *Nitzschia* spp. are the most prevalent. Among cyanobacteria, the filamentous type, such as *Calothrix* spp., *Leptolyngbya* spp. and *Phormidium* spp., is the most common. These blue-green algae are able to form mats on their own or to be part of more diverse communities. The most common genera of green algae seem to vary more between countries than other groups of algae. Nevertheless, they are also cosmopolitan. *Apatococcus*, *Chlorella*, *Chlorosarcinopsis*, *Stichococcus* and *Scenedesmus* are the most frequent genera among all the countries. However, putting some differences in the distribution of the genera aside, it is apparent that the same types of microalgae colonize fountains all around the world. Therefore, a key that allows to identify them could be used internationally by restorers from any country.

Knowing the main types of phototrophic groups present on the monumental fountains helps to better understand the risk related to the biologically induced biodeterioration phenomena.

4.2. Phototrophic Control on the Fountains

Fountains need regular maintenance and restoration work [46] due both to material alterations and to the rapid colonization by phototrophic microorganisms that affects their esthetic value and induces biodeterioration processes.

Parameters such as the pH (an alkaline pH accelerates the algal growth, while an acidic one dissolves carbonatic stones and induces copper staining on stone), the calcium hardness (induces crust formations), the dissolved solids (organic and inorganic, favor biological growth and the clogging of pipes) and the temperature (air and water) are key elements in promoting or not the biodeterioration phenomena. Often, the water is recycled and chemically treated to avoid biological growth [47], but this only helps to postpone the biological development.

Different other strategies [48] can be used in controlling the biological development in the monumental fountains. One is related with the use of antifouling agents aiming to deter the biofilm formation [49–51]. They should be applied on cleaned surfaces in order to postpone biological colonization. These methods are still under investigations for defining practical effectiveness. Another strategy is related to the elimination of the biological growth with the help of biocides [52,53] and other chemical cleaners, such as chlorine. A new trend in controlling the unwanted phototrophic growth is the use of viruses [54]. However, the implementation of a management/maintenance plan (regular checks, periodic tests of water quality, protection from freeze–thaw damage, use of inert control treatments) is preferred, while knowing the phototrophic biodiversity (identified to the genera level by using these keys) helps in monitoring changes in the quality of water and in programming the interventions.

5. Conclusions

These identification keys could be a useful instrument for a better understanding of the phototrophic biodeteriogens present on the artistic fountains and other wet stone surfaces. These keys are indicative, and it is highly recommended to not assign a genera name to the observed microorganisms if the specific features are not observed, keeping the identification to the upper level (e.g., filamentous cyanobacteria with no clearly developed trichomes).

Supplementary Materials: The following are available online at https://www.mdpi.com/article/10.3390/app11188787/s1, Figure S1: Microscopic observations of different samples taken from Sultana Fountain and North "Guitar" Fountain of the Court of the Myrtles, both in the Alhambra complex, Spain, Figure S2: Microscopic observations of different samples taken from Sultana Fountain and the Lions Fountain in the Alhambra complex, Spain, and from Villa la Pietra Fountain from Florence, Italy, Figure S3: Microscopic observations of different samples taken from Villa la Pietra Fountain from Florence and from different fountains from the Alhambra complex, Spain, Figure S4: Microscopic observations of different samples taken from Tacca Fountain from Florence, Italy, and from two fountains from the Alhambra complex in Spain, Figure S5: Microscopic observations of different samples taken from different fountains from the Alhambra complex, Spain, Figure S6: Microscopic observations of some isolated cyanobacteria from monumental fountains: *Aphanocapsa* (**a**), *Aphanothece* (**b**), *Synechococcus* (**c**), *Gloeobacter* (**d**), *Chroococcus* (**e**), and *Gloeocapsa* (**f**). Figure S7: Microscopic observations of some isolated cyanobacteria from monumental fountains: *Leptolyngbya*, that is thinner than *Oscillatoria*, presents sheath and lacks motility (**a**,**b**), *Pseudoanabaena* (**c**), *Pseudophormidium* (**d**), *Calothrix* (**e**,**f**). Figure S8: Microscopic observations of some isolated cyanobacteria from monumental fountains: *Nostoc* (**a**,**b**), *Pleurocapsa* (**c**), *Rivularia* (**d**). Figure S9: Microscopic observations of some isolated cyanobacteria from monumental fountains: *Chlorella* (**a**,**b**), *Scenedesmus* (**c**), *Bracteococcus* (**d**), *Monoraphidium* (**e**), *Dilabifilum* (**f**). Figure S10: Microscopic observations of some isolated cyanobacteria from monumental fountains: *Diatoma* (**a**), *Nitzchia* (**b**), *Achnantes* (**c**), *Surirella* (**d**).

Author Contributions: Conceptualization, F.B.-G., P.S.-C. and O.A.C.; methodology, F.B.-G., P.S.-C., O.A.C.; investigation, F.B.-G., P.S.-C., O.A.C., C.A.-R.; writing—review and editing, O.A.C., F.B.-G., P.S.-C. and C.A.-R.; supervision, O.A.C., F.B.-G. and P.S.-C.; project administration, F.B.-G.; funding acquisition, F.B.-G. All authors have read and agreed to the published version of the manuscript.

Funding: This research was funded by the GOVERNMENT OF ANDALUSIA AND EUROPEAN REGIONAL DEVELOPMENT FUND, grant number A-HUM-279-UGR18 and P18-FR-4477 (FICOARTE regional project), and by the SPANISH MINISTRY OF ECONOMY AND COMPETITIVENESS, grant number PID2019.109713RB.100 (VIRARTE national project).

Institutional Review Board Statement: Not applicable.

Informed Consent Statement: Not applicable.

Acknowledgments: The authors would like to thank the Alhambra and Generalife Patronate (Granada, Spain), the Real Alcázar de Sevilla (Seville, Spain), the Smithsonian Institution (Washington DC, USA) and the Florence Municipality and Villa la Pietra (Florence, Italy) for their assistance in collecting samples, the Ligalismo Foundation for its ongoing dissemination activity, and the Vice-Rectorate for Research and Knowledge Transfer of the University of Granada for their support and the creation of the unit of excellence Ciencia en la Alhambra.

Conflicts of Interest: The authors declare no conflict of interest. The funders had no role in the design of the study; in the collection, analyses, or interpretation of data; in the writing of the manuscript, or in the decision to publish the results.

References

1. Juuti, P.S.; Antoniou, G.P.; Dragoni, W.; El-Gohary, F.; De Feo, G.; Katko, T.S.; Rajala, R.P.; Zheng, X.Y.; Drusiani, R.; Angelakis, A.N. Short Global History of Fountains. *Water* **2015**, *7*, 2314–2348. [CrossRef]
2. Tomaselli, L.; Pietrini, A.M. Alghe e cianobatteri. In *La Biologia Vegetale per i Beni Culturali. Biodeterioramento e Conservazione*; Caneva, G., Nugari, M.P., Salvadori, O., Eds.; Nardini Editore: Firenze, Italy, 2007; Volume 1, pp. 71–77.
3. Gattuso, C.; Cozza, R.; Gattuso, P.; Villella, F. *La Conoscenza Per il Restauro e la Conservazione. Il Ninfeo di Vadue a Carolei e la Fontana Nuova di Lamezia Terme*; Franco Angeli S.R.L.: Milano, Italy, 2012; p. 75.
4. Altieri, A.; Pietrini, A.M.; Ricci, S.; Roccardi, A. Il Controllo del Biodeterioramento. In *Il Restauro della Fontana del Fuga nell'Orto Botanico di Roma*; Micheli, M.P., Tammeo, G., Eds.; Gangemi Editore Spa: Roma, Italy, 2011; pp. 111–117.
5. Villalba Corredor, L.S.; Malagón Forero, A. Biodeterioro de la Fuente de Lavapatas, parque arqueológico de San Agustín-Huila. Colombia. *Ge-Conserv.* **2011**, *2*, 65–80. [CrossRef]
6. Sánchez-Castillo, P.M.; Bolivar-Galiano, F.C. Characterización de comunidades algales epilìticas en fuentes monumentales y aplicacion a la diagnosis del biodeterioro. *Limnetica* **1997**, *13*, 31–46.
7. Not, R.; Miceli, G.; Terranova, F.; Catalisano, A.; Lo Campo, P. Indagini sulla natura biotica delle alterazioni. In *Fontana Pretoria—Studi Per un Progetto di Restauro. Regione Siciliana—Assessorato dei Beni Culturali e Ambientali e della Pubblica Istruzione, Centro Regionale Per la Progettazione e il Restauro e Per le Scienze Naturali ed Applicate ai Beni Culturali, Palermo*; Catalisano, A., Di Natale, R., Gallo, E., Vergara, F., Eds.; Arti Grafiche Siciliane: Palermo, Italy, 1996; pp. 51–64.
8. Hamadeh, S. Splash and Spectacle: The Obsession with Fountains in Eighteenth-Century Istanbul. *Muqarnas Online* **2002**, *19*, 123. [CrossRef]
9. Guillaud, H. *Socio-Cultural Sustainability in Vernacular Architecture. Versus: Heritage for Tomorrow*; Lotti, L., Ed.; Firenze University Press: Firenze, Italy, 2014; pp. 48–55.
10. Delgado-Rodrigues, J. Stone patina. A controversial concept of relevant importance in conservation. In *Proceedings of International Seminar Theory and Practice in Conservation-A Tribute to Cesare Brandi, Lisbon, Portugal, 4–5 May 2006*; Delgado Rodrigues, J., Mimoso, J.-M., Eds.; LNEC Editor: Lisbon, Portugal, 2006; pp. 163–174.
11. Bolivar-Galiano, F.C.; Sánchez-Castillo, P.M. Claves de identificación de microalgas frecuentes en monumentos. *PH: Boletín Del Inst. Andal. Del Patrim. Histórico* **1999**, *7*, 93–98. [CrossRef]
12. Komárek, J.; Anagnostidis, K. Cyanoprokaryota Chroococcales. In *Süsswasserflora von Mitteleuropa*; Ettl, H., Gärtner, G., Heynig, H., Mollenhauer, D., Eds.; Gustav Fischer Verlag, JenaStuttgart-Lübeck-Ulm: Jena, Germany, 1998; p. 548.
13. Biggs, B.J.F.; Kilroy, C. Identification guide to common periphyton in New Zealand streams and rivers. In *Stream Periphyton Monitoring Manual, Prepared for The New Zealand Ministry for the Environment*; NIWA: Christchurch, New Zealand, 2000; pp. 121–210.
14. Komárek, J. Coccoid and Colonial Cyanobacteria. In *Freshwater Algae of North America: Ecology and Classification*; Wehr, J.D., Sheath, R.G., Eds.; Academic Press: Cambridge, MA, USA, 2003; pp. 59–116.
15. B-Komárek, J.; Komárková, J.; Kling, H. Filamentous cyanobacteria. *Freshwater Algae of North America: Ecology and Classification*, Wehr , J.D., Sheath, R.G., Eds.; Academic Press: Cambridge, MA, USA, 2003; 117–196.

16. Komárek, J.; Anagnostidis, K. Cyanoprokaryota. 2. Teil: Oscillatoriales. In *Süsswasserflora von Mitteleuropa*; Büdel, B., Krienitz, L., Gärtner, G., Schagerl, M., Eds.; Springer Spektrum: Heidelberg, Germany, 2005; pp. 1–759.
17. Komárek, J. Cyanoprokaryota. Heterocytous genera. In *Süswasserflora von Mitteleuropa/Freshwater Flora of Central Europe*; Büdel, B., Gärtner, G., Krienitz, L., Schagerl, M., Eds.; Springer Spektrum: Berlin/Heidelberg, Germany, 2013; p. 1130.
18. Komárek, J.; Johansen, J.R. Filamentous Cyanobacteria. In *Freshwater Algae of North America*, 2nd ed; Academic Press: Cambridge, MA, USA, 2015; pp. 135–235. [CrossRef]
19. Komárek, J.; Johansen, J. Coccoid Cyanobacteria. In *Freshwater Algae of North America*, 2nd ed; Academic Press: Cambridge, MA, USA, 2015; pp. 75–133. [CrossRef]
20. Bourrelly, P. Diatomophycées. In *Les Alguesd'eaudouce. Initiation à la Systématique. Tome II: Les Algues Jaunes et Brunes-Chrysophyycées, Phéophycées, Xanthophycées et Diatomées*; Boubée, N., Ed.; Cie: Paris, France, 1968; pp. 259–399.
21. Bolívar-Galiano, F.; Abad-Ruiz, C.; Sánchez-Castillo, P.; Toscano, M.; Romero-Noguera, J. Frequent Microalgae in the Fountains of the Alhambra and Generalife: Identification and Creation of a Culture Collection. *Appl. Sci.* **2020**, *10*, 6603. [CrossRef]
22. Bolívar-Galiano, F.C.; Abad-Ruiz, C.; Yebra, A.; Romero-Noguera, J.; Sánchez-Castillo, P. Chromatic alterations by microalgae at National Mall fountains in Washington DC (USA). In *Proceedings of the 6th International Conference on Heritage and Sustainable Development, Granada, Spain, 12–15 June 2018, Granada, Spain*; EUG Green Lines Institute for Sustainable Development: Barcelos, Portugal, 2018; Volume 2, pp. 1211–1218.
23. Cuzman, O.A.; Ventura, S.; Sili, C.; Mascalchi, C.; Turchetti, T.; D'Acqui, L.P.; Tiano, P. Biodiversity of Phototrophic Biofilms Dwelling on Monumental Fountains. *Microb. Ecol.* **2010**, *60*, 81–95. [CrossRef]
24. Zurita, Y.P.; Cultrone, G.; Castillo, P.S.; Sebastián, E.; Bolívar, F. Microalgae associated with deteriorated stonework of the fountain of Bibatauín in Granada, Spain. *Int. Biodeterior. Biodegrad.* **2005**, *55*, 55–61. [CrossRef]
25. Peraza Zurita, Y. Biodeterioro por Microalgas en Fuentes de Mármol. Ph.D. Thesis, University of Granada Departamento de Pintura, Granada, Spain, 2004.
26. Bolívar-Galiano, F.C.; Sánchez-Castillo, P.M. Biodeterioro del patrimonio artístico por cianobacterias, algas verdes y diatomeas. *Idea PH Boletìn* **1998**, *24*, 52–63. [CrossRef]
27. De Natale, A.; Mele, B.H.; Cennamo, P.; Del Mondo, A.; Petraretti, M.; Pollio, A. Microbial biofilm community structure and composition on the lithic substrates of Herculaneum Suburban Baths. *PLoS ONE* **2020**, *15*, e0232512. [CrossRef] [PubMed]
28. Pietrini, A.M. La microflora fotosintetica. In *Il restauro della Fontana del Fuga nell'Orto Botanico di Roma*; Micheli, M.P., Tammeo, G., Eds.; Gangemi Editore Spa: Roma, Italy, 2011; pp. 107–111.
29. Sarró, M.I.; Garcia, A.M.; Rivalta, V.M.; Moreno, D.A.; Arroyo, I. Biodeterioration of the Lions Fountain at the Alhambra Palace, Granada (Spain). *Build. Environ.* **2006**, *41*, 1811–1820. [CrossRef]
30. Hindáková, A.; Hindác, F. Green algae of five city fountains in Bratislava (Slovakia). *Biol. Bratisl.* **1998**, *53*, 481–493.
31. Nugari, M.P.; Pietrini, A.M. Trevi Fountain: An evaluation of inhibition effect of water-repellents on cyanobacteria and algae. *Int. Biodeterior. Biodegrad.* **1997**, *40*, 247–253. [CrossRef]
32. Gandhi, H.P. Notes on the diatomaceae from Ahmedabad and its environs ? VI. On some diatoms from fountain reservoir of Seth Sarabhai's garden. *Hydrobiology* **1967**, *30*, 248–272. [CrossRef]
33. Chorus, I.; Cavalieri, M. Cyanobacteria and algae. In *Monitoring Bathing Waters—A practical Guide to the Design and Implementation of Assessments and Monitoring Programme*; Bartram, J., Rees, G., Eds.; CRC Press: London, UK, 2000; pp. 219–272. Available online: https://www.who.int/water_sanitation_health/bathing/bathwatchap10.pdf (accessed on 28 August 2021).
34. Trobajo, R.; Mann, D.G. A rapid cleaning method for diatoms. *Diatom Res.* **2019**, *34*, 115–124. [CrossRef]
35. Franchini, W. The Collecting, Cleaning, and Mounting of Diatoms. Available online: https://www.mccrone.com/mm/the-collecting-cleaning-and-mounting-of-diatoms/ (accessed on 30 August 2021).
36. Popović, S.; Nikolić, N.; Jovanović, J.; Predojević, D.; Trbojević, I.; Manić, L.; Subakov Simić, G. Cyanobacterial and algal abundance and biomass in cave biofilms and relation to environmental and biofilm parameters. *Int. J. Speleol.* **2019**, *48*, 49–61. [CrossRef]
37. Bolívar-Galiano, F.C. Diagnosis y Tratamiento del Deterioro por Microalgas en los Palacios Nazaries de la Alhambra. Ph.D. Thesis, Universidad de Granada, Granada, Spain, 1994; pp. 193–218. Available online: https://digibug.ugr.es/ (accessed on 1 September 2021).
38. *Phyco Key, A List of Helpful Texts and Websites with Excellent Images*; Baker, A.L. (Ed.) University of Hampshire: Hampshire, UK; Available online: http://cfb.unh.edu/phycokey/Choices/Text_html/groups/genera.htm (accessed on 1 September 2021).
39. Mihajlovski, A.; Seyer, D.; Benamara, H.; Bousta, F.; Di Martino, P. An overview of techniques for the characterization and quantification of microbial colonization on stone monuments. *Ann. Microbiol.* **2015**, *65*, 1243–1255. [CrossRef]
40. Piñar, G.; Sterflinger, K. Natural sciences at the service of art and cultural heritage: An interdisciplinary area in development and important challenges. *Microb. Biotechnol.* **2021**, *14*, 806–809. [CrossRef]
41. Pandey, D.C.; Mitra, A.K. On the morphology and life-history of a form of gloeocapsa showing some new stages. *Hydrobiologia* **1966**, *27*, 379–384. [CrossRef]
42. Absalón, I.B.; Muñoz-Martín, M.A.; Montejano, G.; Mateo, P. Differences in the Cyanobacterial Community Composition of Biocrusts from the Drylands of Central Mexico. Are There Endemic Species? *Front. Microbiol.* **2019**, *10*, 937. [CrossRef]
43. Charola, A.E.; Wendler, E. An Overview of the Water-Porous Building Materials Interactions. *Restor. Build. Monum.* **2015**, *21*, 55–65. [CrossRef]

44. Gaylarde, C.; Gaylarde, P.M.; Neilan, B. Endolithic Phototrophs in Built and Natural Stone. *Curr. Microbiol.* **2012**, *65*, 183–188. [CrossRef]
45. Cuzman O., A.; Tiano, P.; Ventura, S.; Frediani, P. Biodiversity on Stone Artifacts. In *The Importance of Biological Interactions in the Study of Biodiversity*; IntechOpen: Rijeka, Croatia, 2011; pp. 367–390.
46. Conference on Conservation of Fountains, organized by The National Center for Preservation Technology and Training, in partnership with The Nelson-Atkins Museum of Art and conservator Martin Burke. Kansas City, MO, USA, 10–11 July 2013. Available online: https://www.ncptt.nps.gov/blog/fountain-fundamentals/ (accessed on 10 July 2021).
47. Krueger, R. A new approach to maintaining water features and reducing biological growth. In *Objects Specialty Group Postprints*; Riccardelli, C., Del Re, C., Eds.; The American Institute for Conservation of Historic & Artistic Works: Washington, DC, USA, 2010; Volume 17, pp. 33–39.
48. Salvadori, O.; Charola, A.E. Methods to Prevent Biocolonization and recolonization: An overview of current research for architectural and archaeological heritage. In *Biocolonization of Stone: Control and Preventive Methods Proceedings from the MCI Workshop Series*; Charola, A.E., McNamara, C., Koestler, R.J., Eds.; Smithsonian Institution Scholarly Press: Washington, DC, USA, 2009; pp. 37–50.
49. Lo Schiavo, S.; De Leo, F.; Urzì, C. Present and Future Perspectives for Biocides and Antifouling Products for Stone-Built Cultural Heritage: Ionic Liquids as a Challenging Alternative. *Appl. Sci.* **2020**, *10*, 6568. [CrossRef]
50. Cuzman, O.A.; Camaiti, M.; Sacchi, B.; Tiano, P. Natural antibiofouling agents as new control method for phototrophic biofilms dwelling on monumental stone surfaces. *Int. J. Conserv. Sci.* **2011**, *2*, 3–16.
51. Yebra, D.M.; Kiil, S.; Dam-Johansen, K. Antifouling technology—past, present and future steps towards efficient and environmentally friendly antifouling coatings. *Prog. Org. Coat.* **2004**, *50*, 75–104. [CrossRef]
52. Pandolfi, A.; Capponi, G.; Fazio, G. The restoration of historical fountains. In *Restoring in Italy—Art and Technology in the Activities of the Istituto Superiore per la Conservazione ed Il Restauro*; Gangemi Editore: Rome, Italy, 2012; pp. 87–94.
53. Gaylarde, C.; Morton, L.H.G. Deteriogenic biofilms on buildings and their control: A review. *Biofouling* **1999**, *14*, 59–74. [CrossRef]
54. May, E.; Zamarreño, D.; Hotchkiss, S.; Mitchell, J.; Inkpen, R. Bioremediation of Algal Contamination on Stone. In *Biocolonization of Stone: Control and Preventive Methods Proceedings from the MCI Workshop Series*; Charola, A.E., McNamara, C., Koestler, R.J., Eds.; Smithsonian Institution Scholarly Press: Washington, DC, USA, 2009; pp. 59–70.

Article

In Vitro Antibacterial Activity of Marine Microalgae Extract against *Vibrio harveyi*

Mohd Ridzwan Jusidin [1], Rafidah Othman [1,*], Sitti Raehanah Muhamad Shaleh [1], Fui Fui Ching [1], Shigeharu Senoo [2] and Siti Nur Hazwani Oslan [3]

[1] Borneo Marine Research Institute, Universiti Malaysia Sabah, Kota Kinabalu 88400, Sabah, Malaysia; ridzwan606@gmail.com (M.R.J.); sittirae@ums.edu.my (S.R.M.S.); cfuifui@ums.edu.my (F.F.C.)
[2] UMS-Kindai Aquaculture Development Center, Kindai University, 3153 Shirahama, Wakayama 649-211, Japan; ssenoo@ums.edu.my
[3] Faculty of Food Science and Nutrition, Universiti Malaysia Sabah, Kota Kinabalu 88400, Sabah, Malaysia; snhazwanioslan@ums.edu.my
* Correspondence: rafidaho@ums.edu.my

Featured Application: Potential application of *N. oceanica*, *Isochrysis* sp. and *T. weissflogii* as inhibitory bacteria and probiotics in controlling bacterial diseases in the aquaculture industry.

Abstract: Marine microalgae may produce antibacterial substances. At the exponential phase of growth, four species of marine microalgae were examined for their potential to create secondary metabolites that limit the growth of *Vibrio harveyi*: *Nannochloropsis oceanica*, *Chaetoceros gracilis*, *Isochrysis* sp. and *Thalassiosira weissflogii*. *V. harveyi* is a pathogenic bacteria that can cause severe mortality and loss in aquaculture. Disc diffusion assay and co-culture assay were used to determine antibacterial activity. On TSA % NaCl media, the disc impregnated with microalgae and extracted with ethanol, methanol, saline water, and dimethyl sulfoxide (DMSO) was tested against *V. harveyi* at concentrations of 1.0×10^5, 10^6 and 10^7 CFU mL^{-1}. The disc diffusion assay revealed that *N. oceanica* extracted with ethanol had the largest inhibitory zone against *V. harveyi*. Meanwhile, only *N. oceanica*, *Isochrysis* sp. and *T. weissflogii* reduced the growth of *V. harveyi* (10^5 CFU mL^{-1}) in the co-culture assay ($p < 0.05$). The current findings reveal that the hydrophilic chemicals in microalgae extract have antibiotic activity against the highly virulent *V. harveyi*, which causes vibriosis, a serious disease in farmed fish and aquaculture cultivation around the world.

Keywords: microalgae; antibacterial; *Vibrio harveyi*; *Nannochloropsis oceanica*; *Chaetoceros gracilis*; *Isochrysis* sp.

Citation: Jusidin, M.R.; Othman, R.; Shaleh, S.R.M.; Ching, F.F.; Senoo, S.; Oslan, S.N.H. In Vitro Antibacterial Activity of Marine Microalgae Extract against *Vibrio harveyi*. *Appl. Sci.* **2022**, *12*, 1148. https://doi.org/10.3390/app12031148

Academic Editors: Anna Poli and Valeria Prigione

Received: 20 December 2021
Accepted: 18 January 2022
Published: 22 January 2022

1. Introduction

In recent years, the aquaculture business has grown quickly and is predicted to supply approximately 62% of fish for human demand and consumption by 2030 [1,2]. However, diseases in aquaculture could impact the aquaculture production function by destroying basic resources, lowering the physical output or unit value of a production process, lowering the efficiency of a production process, and ultimately leading to economic losses in the aquaculture sector [2]. Vibriosis is one of the most prevalent bacterial diseases and is claimed to have damaged farmed marine fish in Malaysia, resulting in a USD 7.4 million loss in 1990 [3] and severe economic loss to Asian seabass farmers in 2017 [4].

Vibriosis is associated with infections in fish, such as skin necrosis, ulceration, and scale drops on the abdomen caused by a variety of *Vibrio* spp., including *Vibrio harveyi*, *V. vulnificus*, *V. parahaemolyticus*, *V. alginolyticus*, *V. penaeicida*, and *V. splendidus* [5]. *V. harveyi*, the primary bacterium responsible for catastrophic death in fish farming, has led to massive economic loss [6]. Antibiotics and chemicals are frequently used by farmers to combat harmful organisms. However, they have been used sparingly since they are expensive, non-biodegradable, highly biomagnified, and antibiotic resistance has grown [7]. Antibiotic overuse can result in various environmental problems, including contamination of the culture environment, organism harm, and the development of bacterial resistance that can extend to the food chain [8]. Alternatively, numerous techniques for controlling pathogenic vibriosis have been proposed, including phage therapy, short-chain fatty acid inhibition of bacterial growth, quorum-sensing disruption, probiotics, immunostimulants, vaccinations, and green water [9].

Pathogen management in aquaculture, particularly disease prevention employing herbs and phytochemicals, has garnered a considerable amount of attention in the previous decade [10]. Furthermore, microalgae can produce a variety of bioactive compounds such as carotenoids, polysaccharides, vitamins, and lipids [11]. Recently, microalgae have been frequently utilised in aquaculture as nutritional supplements, immunostimulants to strengthen immune systems, and to increase disease resistance against harmful bacteria [2]. Furthermore, they contain antibacterial and antiviral properties that could be exploited for disease prevention and management in the aquaculture industry [11]. Thus, because the habitat is in a trapped and limited area, such as in fish-rearing tanks or cage culture, the transmission of numerous viral, fungal, parasitic, and bacterial diseases easily occurs among cultured fish, which can further cause co-infections and mortality in cultured fish [12,13]. It has been found that adding microalgae to farmed fish reduces the bacterial load of larval rearing systems by reducing the number of opportunistic bacteria [14].

Furthermore, the growing desire for more ecologically friendly disease control strategies has prompted academics to investigate alternate ways with little negative effects. Thus, the primary goal of this work was to investigate the antibacterial activity effects of several marine microalgae cultures, including *Nannochloropsis oceanica*, *Chaetoceros gracilis*, *Isochrysis* sp. and *Thalassiosira weissflogii*, against *V. harveyi* at different cell viability levels.

2. Materials and Methods

2.1. Cultures and Preparation of Microalgae Extract

The species of unicellular marine microalgae were tested. *N. oceanica*, *C. gracilis*, *Isochrysis* sp. and *T. weissflogii* microalgae strains were collected from the Culture Collection of the Live Feed Culture Laboratory, Borneo Marine Research Institute, University Malaysia Sabah. Each microalgae culture was produced and maintained in 250 mL Erlenmeyer flasks with 100 mL Guillard's F/2 media [15]. The microalgae culture was reactivated as an inoculum for four days and the starting culture concentration was 1×10^5 cells mL^{-1}. The cultivation conditions were set at 25 ± 1 °C with 24 h of continuous illumination at 1000 μmol photon m^{-2} s^{-1} given by cool white fluorescent lights. The microalgae inoculum culture was put into 500 mL Erlenmeyer flasks with 200 mL of Walney's media [16]. After 10 days in culture, the cultivation reached the exponential phase of growth and cell density (1×10^7 cells mL^{-1}). The cell viability of cell microalgae was evaluated using a Malassez haemocytometer on a daily basis. By plating 50 μL samples to tryptic soy agar (TSA) with 2% (w/v) NaCl and Thiosulfate Citrate Bile Salt Sucrose (TCBS) agar, the viability of cell microalgae culture was tested for contamination. The microalgae culture was afterwards recovered and harvested by centrifugation at low speed (2000× g) for 10 min at 15 °C, and the algal pellet was filtered through Whatman no. 1 filter paper [17]. The extraction of microalgae biomass was carried out with four different solvents: methanol, ethanol, Dimethyl sulfoxide (DMSO), and saline water, with a ratio of 0.1 g of algal biomass for 1.0 mL of solvent. Every microalgae extract was fixed into 10^6 cells mL^{-1}. Before further usage, the crude extracts were promptly refrigerated at 4 °C.

2.2. Preparation of Vibrio harveyi Inoculums

Vibrio strains were collected from the Live Feed Culture Laboratory Culture Collection at the Borneo Marine Research Institute, University Malaysia Sabah. *V. harveyi* strains were obtained from an epidemic of vibriosis in Asian seabass, *Lates calcarifer* (Bloch). *V. harveyi* stock culture was subcultured into TCBS agar plates and incubated overnight at 28 °C.

After 24 h of incubation, pure single colonies were selected and subcultured in 1.5 percent NaCl Tryptic Soy Broth (TSB). The bacterium was then cultured for 24 h at an incubator shaker. Then, 1 mL of inoculum was transferred to a 1.5 mL microcentrifuge tube and centrifuged for 5 min at 2000 rpm (10,000× *g*). After discarding the supernatant, 700 μL of TSB containing 1.5% NaCl was added to the tube and vortexed to mix the bacterial pellet with the TSB. To obtain varying amounts of inoculum, appropriate dilutions were made, and the viable cell count was validated using a spread plate test using a serial dilution of the bacterial suspension.

2.3. Disc Diffusion Antibacterial Assay

The various microalgae extractions were tested for their antibacterial activity against the pathogenic bacteria strains of *V. harveyi*. A modified agar disc diffusion assay method was used to investigate the antibacterial activity of algal extracts [18]. Petri dishes containing TSA and 2% (*w*/*v*) NaCl were seeded with *V. harveyi* inoculum at three different concentrations: 10^5 CFU mL^{-1}, 10^6 CFU mL^{-1}, and 10^7 CFU mL^{-1}. To test the extracts' activity, sterile filter paper discs (6 mm) were impregnated with 20 μL of the various algal extracts. The discs were allowed to dry at room temperature before being placed on test plates inoculated with *V. harveyi*. As a control, discs with the same volume of extractants (20 μL) were made. The plates were incubated for 48 h at 35 °C. Extracts containing antibacterial components produced distinct, clear, and circular zones of inhibition around the filter discs, and this positive activity was quantified by measuring the growth inhibition zone (mm) surrounding the discs after 24 h and scoring them based on the diameter of the inhibition zone [19].

2.4. Co-Culture Antibacterial Assay

Cultures of four microalgae species, *N. oceanica*, *C. gracilis*, *Isochrysis* sp. and *T. weissflogii*, were utilised at 10^6 CFU mL^{-1} concentration to evaluate their potential to inhibit the various viability of cell growth of *V. harveyi* by incubating this bacteria in these microalgae cultures (co-culture). The concentration of *V. harveyi* used in this co-culture experiment was determined by the disc diffusion assay result that indicated the broadest inhibitory zone. Co-culture assay samples were collected at 0, 12, 24, 48, 72, 96, and 120 h after the experiment began, and 50 μL of 10-fold dilutions were spread on TCBS agar plates to assess the bacterial content in the samples. All studies were performed in triplicate, and each microalga studied, as well as *V. harveyi*, received a control treatment.

2.5. Statistical Analysis

All data for the concentration of *V. harveyi* at each point of time for co-culture assay were statistically analyzed by one-way ANOVA using software SPSS (version 22), and graphs have been plotted using Microsoft Excel.

3. Results

3.1. Disc Diffusion Antibacterial Assay

According to the findings of this investigation, four microalgae extracts showed varying degrees of inhibitory impact against *V. harveyi*, as shown in Table 1, and the extraction had broad-spectrum activity against pathogenic bacteria. On the other hand, ethanolic extracts inhibited all microalgae with the broadest (+++) range of inhibition against the lowest concentration of *V. harveyi* at 10^5 CFU mL^{-1}. However, at the maximum concentration of *V. harveyi* (10^7 CFU mL^{-1}), *N. oceanica*, *Isochrysis* sp. and *T. weissflogii* extracts displayed a moderate (++) spectrum of inhibition, whilst *C. gracilis* showed low (+) inhibition. Interestingly, the ethanolic extract of *N. oceanica* demonstrated the broadest (+++) inhibitory zone against all *V. harveyi* concentrations tested; 10^5, 10^6, and 10^7 CFU mL^{-1}.

Table 1. Antibacterial susceptibility test of the microbial extract against *V. harveyi*.

Concentration of *V. harvey*	Extractant	Diameter of the Inhibition Zone (D) in mm by Marine Microalgae Extract (10^6 Cell/mL) and Control (Extractant) after 24 h Incubation with *V. harveyi*				
		N. oceanica	*T. weissflogii*	*C. gracilis*	*Isochrysis* sp.	*Control (Extractant)*
10^5 CFU/mL	Ethanol	+++	+++	+++	+++	–
	Methanol	++	++	++	++	–
	Saline water	++	+	++	++	–
	DMSO	++	+	+	+	–
10^6 CFU/mL	Ethanol	+++	++	++	+++	–
	Methanol	++	++	+	++	–
	Saline water	++	+	+	++	–
	DMSO	+	+	+	+	–
10^7 CFU/mL	Ethanol	+++	++	+	++	–
	Methanol	+	+	+	+	–
	Saline water	+	+	+	+	–
	DMSO	+	+	+	+	–

Note: (−): No activity, (+): D < 6 mm, (++): 6 < D < 8.5 mm, (+++): D > 8.5 mm. D: Diameter of the inhibition zone in milimeters.

3.2. Co-Culture Antibacterial Assay

In the co-culture assay, three (3) microalgae species—co-culture assays of *N. oceanica* (Figure 1a), *T. weissflogii* (Figure 1b), and *Isochrysis* sp. (Figure 1d)—demonstrated minimal antibacterial activity by preventing significant ($p < 0.05$) proliferation of *V. harveyi* at different time intervals when compared to the control group culture of *V. harveyi* with no microalgae.

In the co-culture *V. harveyi* with *N. oceanica* (Figure 1a), the concentration of *V. harveyi* was significantly lower ($p < 0.05$) at 24 to 48 h incubation as compared to control. However, there was no significant difference ($p > 0.05$) in *V. harveyi* concentration at 72, 96, and 120 h incubation.

For the co-culture *V. harveyi* with *T. weissflogii* (Figure 1b), the concentration of *V. harveyi* was significantly lower ($p < 0.05$) at 24, 48, and 72 h incubation, followed by no significant difference ($p > 0.05$) at 96 and 120 h incubation as compared to control.

Meanwhile, the co-culture *V. harveyi* with *Isochrysis* sp in Figure 1d demonstrated a significantly lower ($p < 0.05$) concentration of *V. harveyi* at 72, 96, and 120 h, as compared to the control. In comparison to the control, co-culture *V. harveyi* with *C. gracilis* (Figure 1c) exhibited no significant difference ($p > 0.05$) in *V. harveyi* concentration. In fact, no bibliographic data on the antibacterial activity of *C. gracilis* described in the prior study is available. The growth of *V. harveyi* strains increased steadily with time up to 48 h until reaching a plateaued condition and subsequently decreased insignificantly ($p > 0.05$) at the end of the experiment in the control group culture with no microalgae added.

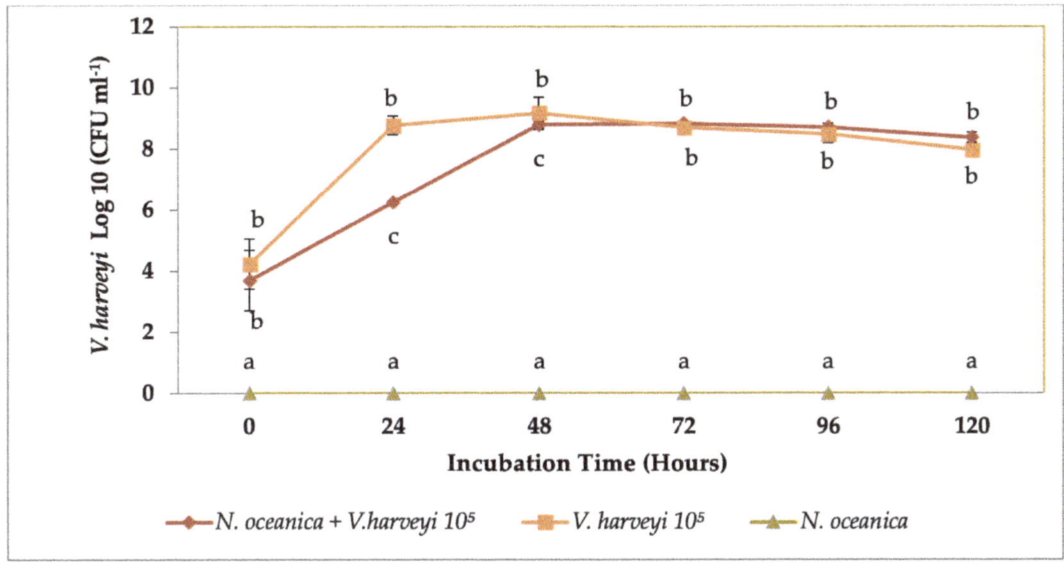

(a)

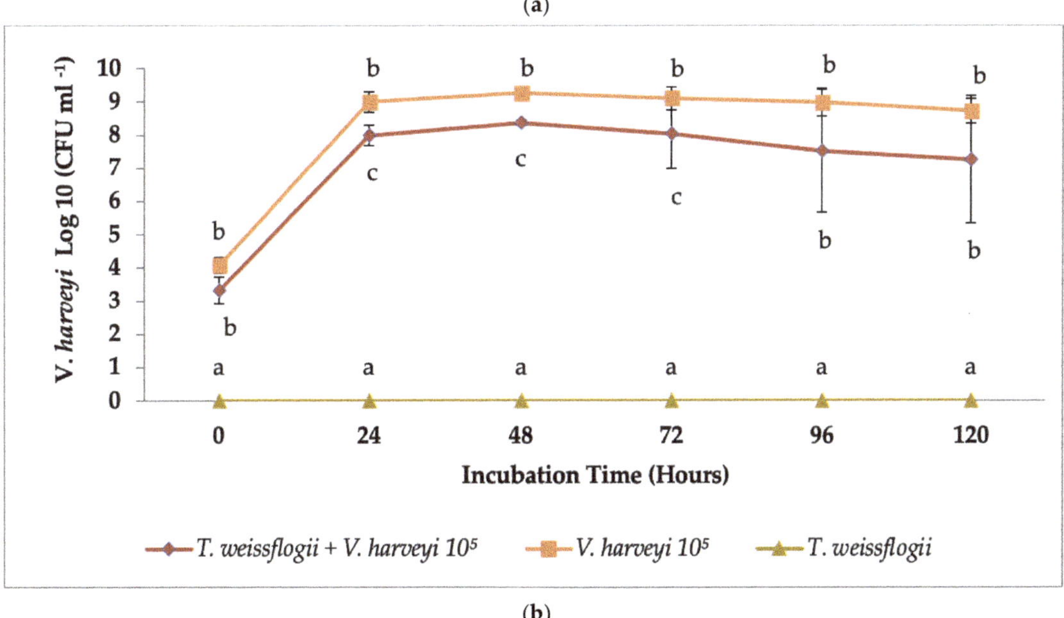

(b)

Figure 1. *Cont.*

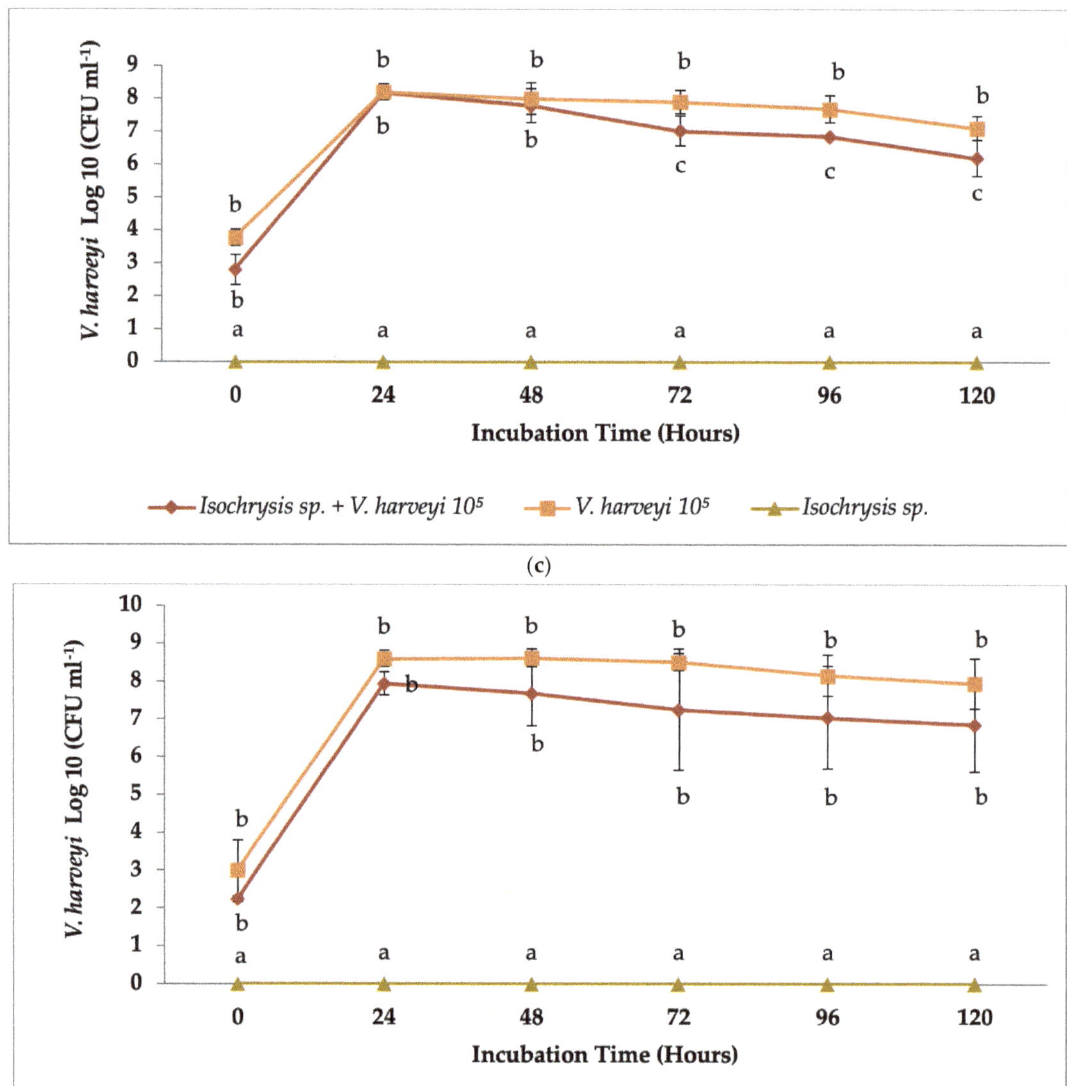

Figure 1. (**a**) *V. harveyi* concentration was significantly lower ($p < 0.05$) at 24–48 h incubation in co-culture assay with *N. oceanica* as compared to control *V. harveyi* with no microalgae added. (**b**) *V. harveyi* concentration was significantly lower ($p < 0.05$) at 24 to 72 h incubation in co-culture assay with *T. weissflogii* as compared to control *V. harveyi* with no microalgae added. (**c**) *V. harveyi* concentration was significantly lower ($p < 0.05$) at 72 to 120 h incubation in co-culture assay with *Isochrysis* sp as compared to control *V. harveyi* with no microalgae added. (**d**) *V. harveyi* concentration showed no significant difference ($p > 0.05$) in co-culture assay with *C. gracilis* as compared to control *V. harveyi* with no microalgae added. Note: Values are presented as mean ±SD. Mean values of *V. harveyi* within the same time (hours) with different lowercase letters are significantly different ($p < 0.05$).

4. Discussion

Many bioactive and pharmacologically active chemicals, particularly antibacterial compounds, are potentially produced by marine microalgae [6,20,21]. Furthermore, microalgae are gaining popularity due to their ability to create bioactive metabolites with anticancer, anti-inflammatory, antibacterial, and antioxidant effects [11,22]. Previous studies have suggested that fatty acids [23], terpenoids, carbohydrates [24], peptides, polysaccharides, and alkaloids are responsible for antibacterial activity in microalgae [25]. Novel antibacterial substances were also discovered in microalgae, with *Coccomyxa onubensis* fatty acid extracts inhibiting *E.coli* and *P. mirabilis* [26]. Furthermore, microalgae have been increasingly treated with other compounds such as cyanovirin, oleic acid, linoleic acid, palmitoleic acid, -carotene, or phycocyanin, which have antioxidant or anti-inflammatory properties, as well as antimicrobial activity, such as against *Staphylococcus aureus* and methicillin-resistant *Staphylococcus aureus* (MRSA) [22,23].

Table 1 shows variable degrees of inhibition on several species of microalgae extract against *V. harveyi* from this investigation. As a result, the generation of antibiotics is heavily reliant on microalgal species [27]. Furthermore, the availability of antibiotic agents can vary greatly between various species within the same class or within a single species, depending on which ecotypes are adapted to certain conditions [27,28]. The green microalga *Dunaliella* sp. isolated from severely polluted waters, for example, was found to be more active against bacteria than ecotypes obtained from less polluted waters [29]. Recently, microalgae of the genus *Nannochloropsis* have been shown to be rich in polyunsaturated fatty acid compounds (PUFAs), carotenoids, polyphenols, and vitamins, and have previously been used in aquaculture [30], while the microalgae strain of *Isochrysis* sp. was shown to be high in fucoxanthin and DHA [31]. *T. weissflogii*, on the other hand, has antibacterial properties due to its high presence of substances such as betain, lipids, phospolipid, polyunsaturated fatty acid, fucoxanthin (FX), and eicosapentaenoic acid (EPA) [32]. *C. gracilis* showed a limited inhibited zone as compared to other microalgae extracts, particularly at higher concentrations of *V. harveyi*. This could be because the compound composition of *C. gracilis* is minimal and contains only non-essential amino acids [33].

Apart from microalgae species, the presence of antibacterial chemicals in microalgae extracts is also strongly reliant on the extraction process and solvent utilised. According to the findings of this investigation, ethanol is the best extractant since it demonstrated the broadest inhibition against *V. harveyi* in the disc diffusion method. Based on prior research, ethanol was also used to extract the thalli of *Gracilaria fisheri* from India, the extract of which exhibited a high inhibitory effect against *V. harveyi* [34]. An ethanol extract of *Gracilaria corticata* from India was found to be highly active against *V. cholerae* and *V. parahaemolyticus*, but less active against *Pseudomonas aeruginosa* and *Shigella flexneri* [35]. Another study on ethanolic extract of *Spiriluna* also demonstrated more effective against two species of *Vibrio* [36]. These findings suggest that molecules with antibacterial action in microalgae are generally hydrophobic and might even be extracted more easily with organic solvents [27]. This could indicate that ethanol is a competent extraction method capable of extracting lipids with a specific set of chemicals present during extraction.

The methanolic extract is a well-established and reported method to isolate active antimicrobial components from microalgae [37,38]. In another study, methanol was found to be the best organic solvent to obtain bioactive compounds from *Dunaliella* sp. [39]. The methanolic extract of freshwater microalgae exhibited antibacterial activity against important human bacterial pathogens [40]. Methanolic extract in the current study inhibited *V. harveyi* moderately, whereas the antibiotic synergism with the methanolic extract of *C. vulgaris* exerted enhanced anti-bacterial activity in *E. coli*. [41]. Therefore, the selection of a solvent is also important for the further utilisation of post-extracted microalgal biomass. According to Navarro et al., 2017 [26], it was suggested that fatty acids could be involved in the antibacterial activity due to increased activity from green microalga extracts obtained with non-polar solvents compared to polar solvents.

In the present study, microalgae extraction by saline water and DMSO demonstrated a comparable less effective antibacterial activity against *V. harveyi* based on the concentration of *V. harveyi*. Nevertheless, it was reported that the extraction of antimicrobial compounds from *Scenedesmus subspicatus* using DMSO inhibited the growth of *Klebsiella pneumoniae* and *Escherichia coli* [42]. As observed in previous studies, water extracts of *Dunaliella* sp. had lower bioactive properties than the ones obtained from organic solvents [43]. Hence, the selection of a solvent is important for optimisation of the extraction process and generation of its bioactive compounds, which have diverse applications not only in aquaculture but also against human pathogens. According to Pradhan et al., 2012 [36] no antimicrobial activity was detected in the aqueous extracts, which was probably because of the low polar nature of the active components.

The number of *V. harveyi* increased exponentially, especially in the first 24 h during the experimental period, without the presence of microalgae cells (control) demonstrating that the bacteria cells were able to utilise the growth medium of the microalgae cultures. The inhibition of *V. harveyi* in co-culture assay with *N. oceanica*, *Isochrysis* sp. and *T. weissflogii* indicated the production of antibacterial compounds by these microalgae cells. Previous research has shown that *Isochrysis* sp. and *Nannochloropsis* exhibit antibacterial action against the majority of vibriosis pathogens, including *V. alginolyticus*, *V. lentus*, *V. splendidus*, *V. scophthalmi*, *V. parahaemolyticus*, and *V. anguillarum* [44]. The microalgae species *Nannochloropsis* sp. and *Isochrysis* sp. have been demonstrated to synthesise short-chain fatty acids, which are implicated in antibacterial activity [45,46]. Thus, microalgae biomass is regarded as a source of valuable chemical elements that can be used in animals, particularly farmed fish, as pharmaceuticals in aquaculture industries.

Another element that could explain these discrepancies is that antibacterial activity is linked to the culture conditions and growth phase of the microalgae cultures. Although all microalgae species were at the same concentration in all trials in this study, the growth phase for each microalgae culture could have been different and may have influenced the results. Previous studies suggested that modifying the culture conditions of green microalgae exhibited differences in antibacterial activity [47–49], whereas modifications on microalgae culture conditions could be stimulated to produce secondary metabolites with antibacterial activity, as well as potentially larger quantities of these secondary metabolites [50]. This may include modifications on media composition, pH, light, and temperature [51,52]. As an example, the highest light intensity (4800 lux) exposure effectively induced the production of antibacterial compounds by *Dunaliella* sp. and suggested that the higher antimicrobial effect was related to the content of bioactive compound produced in stress conditions [39]. Other than the culture environment, the period of microalgae cultivation also plays an important parameter in producing bioactive compounds [39,53]. Therefore, the culture environment could be artificially manipulated in order to allow for increased production of antibacterial compounds in microalgae.

5. Conclusions

Out of four species of microalgae, only *N. oceanica*, *Isochrysis* sp. and *T. weissflogii* demonstrated a significant (($p < 0.05$) inhibitory effect against *V. harveyi* in an in vitro co-culture assay. This indicated a positive effect of the addition of these microalgae in the rearing of fish larvae and implicated the production of antibacterial compounds. Ethanolic extract from *N. oceanica*, *Isochrysis* sp. and *T. weissflogii* demonstrated promising antibacterial activity against *V. harveyi*. Meanwhile, *C. gracilis* inhibited the development of *V. harveyi* with minimal inhibition but no significant difference ($p > 0.05$) in the co-culture assay and disc diffusion method, indicating that it had less potential to limit the growth of *V. harveyi*. Therefore, in the aquaculture industry, *N. oceanica*, *Isochrysis* sp. and *T. weissflogii* act as potential inhibitory bacteria and probiotics in controlling the disease. More comprehensive studies are recommended to optimize its cultivation conditions, the extraction process, and the purification of its antibacterial compounds to unleash their potential.

Author Contributions: Conceptualization, R.O., S.R.M.S. and M.R.J.; methodology, R.O. and M.R.J.; validation, S.R.M.S. and R.O.; formal analysis, M.R.J.; investigation, M.R.J.; resources, F.F.C. and S.R.M.S.; data curation, M.R.J. and R.O.; writing—original draft preparation, M.R.J., R.O. and S.N.H.O.; writing—review and editing, R.O., S.N.H.O. and S.S.; visualization, M.R.J. and F.F.C.; supervision, R.O. All authors have read and agreed to the published version of the manuscript.

Funding: This project was funded by the Minister of Higher Education Malaysia under the Fundamental Research Grant Scheme FRGS0469-2017.

Institutional Review Board Statement: Not applicable.

Informed Consent Statement: Not applicable.

Data Availability Statement: The datasets generated during and/or analyzed during the current study are available from the corresponding author on reasonable request.

Acknowledgments: This project was funded by the Minister of Higher Education Malaysia under the Fundamental Research Grant Scheme FRGS0469-2017. The publication fee was provided by the Research Management Centre, Universiti Malaysia Sabah.

Conflicts of Interest: The authors declare no conflict of interest.

References

1. Pauly, D.; Zeller, D. Comments on FAOs State of World Fisheries and Aquaculture (SOFIA 2016). *Mar. Policy* **2017**, *77*, 176–181. [CrossRef]
2. Ma, K.; Bao, Q.; Wu, Y.; Chen, S.; Zhao, S.; Wu, H.; Fan, J. Evaluation of Microalgae as Immunostimulants and Recombinant Vaccines for Diseases Prevention and Control in Aquaculture. *Front. Bioeng. Biotechnol.* **2020**, *8*, 1331. [CrossRef] [PubMed]
3. Wei, L.S.; Wee, W. Diseases in aquaculture. *Res. J. Ani. Vet. Sci.* **2014**, *7*, 1–6.
4. Mohd Yazid, S.H.; Mohd Daud, H.; Azmai, M.N.A.; Mohamad, N.; Mohd Nor, N. Estimating the Economic Loss Due to Vibriosis in Net-Cage Cultured Asian Seabass (Lates calcarifer): Evidence from the East Coast of Peninsular Malaysia. *Front. Vet. Sci.* **2021**, *8*, 1111. [CrossRef]
5. Ina-Salwany, M.Y.; Al-Saari, N.; Mohamad, A.; Mursidi, F.A.; Mohd-Aris, A.; Amal, M.N.A.; Kasai, H.; Mino, S.; Sawabe, T.; Zamri-Saad, M. Vibriosis in Fish: A Review on Disease Development and Prevention. *J. Aquat. Anim. Health* **2019**, *31*, 3–22. [CrossRef]
6. Farisa, M.Y.; Namaskara, K.E.; Yusuf, M.B.; Desrina. Antibacterial Potention of Extract of Rotifers Fed with Different Microalgae to Control *Vibrio harveyi*. *IOP Conf. Ser. Earth Environ. Sci.* **2019**, *246*, 012058. [CrossRef]
7. Aftabuddin, S.; Siddique, M.A.M.; Habib, A.; Akter, S.; Hossen, S.; Tanchangya, P.; Abdullah, M.A.l. Effects of seaweeds extract on growth, survival, antibacterial activities, and immune responses of *Penaeus monodon* against *Vibrio parahaemolyticus*. *Ital. J. Anim. Sci.* **2021**, *20*, 243–255. [CrossRef]
8. Manyi-Loh, C.; Mamphweli, S.; Meyer, E.; Okoh, A. Antibiotic Use in Agriculture and Its Consequential Resistance in Environmental Sources: Potential Public Health Implications. *Molecules* **2018**, *23*, 795. [CrossRef] [PubMed]
9. Molina-Cárdenas, C.A.; Sánchez-Saavedra, M.P.; Lizárraga-Partida, M.L. Inhibition of pathogenic *Vibrio* by the microalgae *Isochrysis galbana*. *J. Appl. Phycol.* **2014**, *26*, 2347–2355. [CrossRef]
10. Reverter, M.; Bontemps, N.T.; Sasal, P.; Saulnier, D. Use of medicinal plants in aquaculture. In *Diagnosis and Control of Diseases of Fish and Shellfish*; Austin, B., Newaj-Fyzul, A., Eds.; Wiley: Hoboken, NJ, USA, 2017; pp. 223–261.
11. Khavari, F.; Saidijam, M.; Taheri, M.; Nouri, F. Microalgae: Therapeutic potentials and applications. *Mol. Biol. Rep.* **2021**, *48*, 4757–4765. [CrossRef]
12. Abdullah, A.; Ramli, R.; Ridzuan, M.S.M.; Murni, M.; Hashim, S.; Sudirwan, F.; Abdullah, S.Z.; Mansor, N.N.; Amira, S.; Zamri-Saad, M.; et al. The presence of Vibrionaceae, Betanodavirus and Iridovirus in marine cage-cultured fish: Role of fish size, water physicochemical parameters and relationships among the pathogens. *Aquac. Rep.* **2017**, *7*, 57–65. [CrossRef]
13. Mohamad, N.; Mustafa, M.; Amal, M.N.A.; Saad, M.Z.; Md-Yasin, I.S.; Al-saari, N. Environmental factors associated with the presence of Vibrionaceae in tropical cage-cultured marine fishes. *J. Aquat. Anim. Health* **2019**, *31*, 154–167. [CrossRef]
14. Salvesen, I.; Reitan, K.I.; Skjermo, J.; Oie, G. Microbial environments in marine larviculture: Impacts of algal growth rates on the bacterial load in six microalgae. *Aquac. Int.* **2000**, *8*, 275–287. [CrossRef]
15. Guillard, R.R.; Ryther, J.H. Studies on marine planktonic diatoms. I. Cyclotella nana Hustedt and Detonula confervacaea (Cleve) Gran. *Can. J. Microbiol.* **1962**, *8*, 229–239. [CrossRef]
16. Sasireka, G.; Muthuvelayudham, R. Effect of Salinity and Iron Stressed on Growth and Lipid Accumulation in Skeletonema costatum for Biodiesel Production. *Res. J. Chem. Sci.* **2015**, *5*, 69–72.
17. Gonzalez, D.V.A.; Platas, G.; Basilio, A. Screening of antimicrobial activities in red, green and brown macroalgae from Gran Canaria (Canary Islands, Spain). *Int. Microbiol.* **2001**, *4*, 35–40.
18. Bauer, A.N.; Kirby, W.M.M.; Sherries, J.C.; Truck, M. Antibiotic susceptibility testing by a standardized single disk method. *Am. J. Clin. Pathol.* **1966**, *45*, 493–496. [CrossRef] [PubMed]

19. Corona, E.; Fernandez-Acero, J.; Bartual, A. Screening study for antibacterial activity from marine and freshwater microalgae. *Int. J. Pharm. Bio Sci.* **2017**, *8*, 189–194.
20. Prasetiya, F.S.; Sunarto, S.; Bachtiar, E.; Mochamad, U.K.A.; Nathanael, B.; Pambudi, A.C.; Lestari, A.D.; Astuty, S.; Mouget, J.-L. Effect of the blue pigment produced by the tropical diatom *Haslea nusantara* on marine organisms from different trophic levels and its bioactivity. *Aquac. Rep.* **2020**, *17*, 100389. [CrossRef]
21. Soto-Rodriguez, S.A.; Lopez-Vela, M.; Soto, M.N. 2021 Inhibitory effect of marine microalgae used in shrimp hatcheries on Vibrio parahaemolyticus responsible for acute hepatopancreatic necrosis disease. *Aquac. Res.* **2021**, 1–11.
22. Barone, M.E.; Murphy, E.; Parkes, R.; Fleming, G.T.A.; Campanile, F.; Thomas, O.P.; Touzet, N. Antibacterial Activity and Amphidinol Profiling of the Marine Dinoflagellate *Amphidinium carterae* (Subclade III). *Int. J. Mol. Sci.* **2021**, *22*, 12196. [CrossRef]
23. Desbois, A.P.; Mearns-Spragg, A.; Smith, V.J. A fatty acid from the diatom Phaeodactylum tricornutum is antibacterial against diverse bacteria including multi-resistant *Staphylococcus aureus* (MRSA). *Mar. Biotechnol.* **2009**, *11*, 45–52. [CrossRef]
24. Duff, D.C.B.; Bruce, D.L. The antibacterial activity of marine planktonic algae. *Can. J. Microbiol.* **1966**, *12*, 877–884. [CrossRef] [PubMed]
25. Borowitzka, M.A. Microalgae as sources of pharmaceuticals and other biologically active compounds. *J. Appl. Phycol.* **1995**, *7*, 3–15. [CrossRef]
26. Navarro, F.; Forján, E.; Vázquez, M.; Toimil, A.; Montero, Z.; Ruiz-Domínguez, M.D.C.; Garbayo, I.; Castaño, M.A.; Vilchez, C.; Vega, J.M. Antimicrobial activity of the acidophilic eukaryotic microalga *Coccomyxa onubensis*. *Phycol Res.* **2017**, *65*, 38–43. [CrossRef]
27. Falaise, C.; François, C.; Travers, M.A.; Morga, B.; Haure, J.; Tremblay, R.; Turcotte, F.; Pasetto, P.; Gastineau, R.; Hardivillier, Y.; et al. Antimicrobial Compounds from Eukaryotic Microalgae against Human Pathogens and Diseases in Aquaculture. *Mar. Drugs* **2016**, *14*, 159. [CrossRef]
28. Pawlik-Skowrońska, B. Resistance, accumulation and allocation of zinc in two ecotypes of the green alga *Stigeoclonium tenue* Kütz. coming from habitats of different heavy metal concentrations. *Aquat. Bot.* **2003**, *75*, 189–198. [CrossRef]
29. Lustigman, B. Comparison of antibiotic production from four ecotypes of the marine alga, *Dunaliella Bull. Environ. Contam. Toxicol.* **1988**, *40*, 18–22. [CrossRef] [PubMed]
30. Zanella, L.; Vianello, F. Microalgae of the genus *Nannochloropsis*: Chemical composition and functional implications for human nutrition. *J. Funct. Foods* **2020**, *68*, 103919. [CrossRef]
31. Sun, Z.; Wang, X.; Liu, J. Screening of *Isochrysis* strains for simultaneous production of docosahexaenoic acid and fucoxanthin. *Algal Res.* **2019**, *41*, 101545. [CrossRef]
32. Marella, T.K.; Tiwari, A. Marine diatom *Thalassiosira weissflogii* based biorefinery for co-production of eicosapentaenoic acid and fucoxanthin. *Bioresour. Technol.* **2020**, *307*, 123245. [CrossRef]
33. Merchie, G.; Lavens, P.; Sorgeloos, P. Optimization of dietary vitamin C in fish and crustacean larvae: A review. *Aquaculture* **1997**, *155*, 165–181. [CrossRef]
34. Kanjana, K.; Radtanatip, T.; Asuvapongpatana, S.; Withyachumnarnkul, B.; Wongprasert, K. Solvent extracts of the red seaweed *Gracilaria fisheri* prevent *Vibrio harveyi* infections in the black tiger shrimp *Penaeus monodon*, *Fish. Shellfish Immunol.* **2011**, *30*, 389–396. [CrossRef]
35. Padmakumar, K.; Ayyakkannu, K. Seasonal variation of antibacterial and antifungal activities of marine algae from southern coast of India. *Bot. Mar.* **1997**, *40*, 507–515. [CrossRef]
36. Pradhan, J.; Das, B.K.; Sahu, S.; Marhual, N.P.; Swain, A.K.; Mishra, B.K.; Eknath, A.E. Traditional antibacterial activity of fresh water microalga *Spirulina platensis* to aquatic pathogens. *Aquac. Res.* **2012**, *43*, 1287–1295. [CrossRef]
37. Patil, L.; Kaliwal, B.B. Microalga *Scenedesmus bajacalifornicus* BBKLP-07, a new source of bioactive compounds with in vitro pharmacological applications. *Bioprocess Biosyst. Eng.* **2019**, *42*, 979–994. [CrossRef] [PubMed]
38. Zea-Obando, C.; Tunin Ley, A.; Turquet, J.; Culioli, G.; Briand, J.F.; Bazire, A.; Réhel, K.; Faÿ, F.; Linossier, I. Anti-bacterial adhesion activity of tropical microalgae extracts. *Molecules* **2018**, *23*, 2180. [CrossRef]
39. Kilic, N.K.; Erdem, K.; Donmez, G. Bioactive compounds produced by *Dunaliella* species, antimicrobial effects and optimization of the efficiency. *Turk. J. Fish. Aquat. Sci.* **2018**, *19*, 923–933.
40. Prakash, J.W.; Antonisamy, J.M.; Jeeve, S. Antimicrobial activity of certain fresh water microalgae from Thamirabarani River, Tamil Nadu, South India. *Asian Pac. J. Tro. Biomed.* **2011**, *1*, S170–S173. [CrossRef]
41. Pradhan, B.; Patra, S.; Dash, S.R.; Nayak, R.; Behera, C.; Jena, M. Evaluation of the anti-bacterial activity of methanolic extract of *Chlorella vulgaris* Beyerinck [Beijerinck] with special reference to antioxidant modulation. *Futur. J. Pharm. Sci.* **2021**, *7*, 17. [CrossRef]
42. Dantas, D.M.D.M.; Oliveira, C.Y.B.D.; Costa, R.M.P.B.; Carneiro-da-Cunha, M.D.G.; Gálvez, A.O.; Bezerra, R.D.S. Evaluation of antioxidant and antibacterial capacity of green microalgae *Scenedesmus subspicatus*. *Food Sci. Technol. Int.* **2019**, *25*, 318–326. [CrossRef] [PubMed]
43. Herrero, M.; Jaime, L.; Martin-Alvarez, J.M.; Cifuentes, A.; Ibanez, E. Optimization of the Extraction of Antioxidants from *Dunaliella salina* Microalga by Pressurized Liquids. *J. Agric. Food Chem.* **2006**, *54*, 5597–5603. [CrossRef] [PubMed]
44. Kokou, F.; Makridis, P.; Kentouri, M.; Divanach, P. Antibacterial activity in microalgae cultures. *Aquac. Res.* **2012**, *43*, 1520–1527. [CrossRef]

45. Roncarati, A.; Meluzzi, A.; Acciari, S.; Tallarico, N.; Melotti, P. Fatty acid composition of different microalgae strains (Nannochloropsis sp., Nannochloropsis oculata (Droop) Hibberd, *Nannochloropsis atomus* Butcher and *Isochrysis* sp. according to the culture phase and the carbon dioxide concentration. *J. World Aquac. Soc.* **2004**, *35*, 401–411. [CrossRef]
46. Defoirdt, T.; Boon, N.; Sorgeloos, P.; Verstraete, W.; Bossier, P. Alternatives to antibiotics to control bacterial infections: Luminescent vibriosis in aquaculture as an example. *Trends Biotechol.* **2007**, *25*, 472–479. [CrossRef]
47. Elkomy, R.; Ibraheem, I.B.M.; Shreadah, M.; Mohammed, R. Optimal conditions for antimicrobial activity production from two microalgae *Chlorella marina* and *Navicula f. delicatula*. *J. Pure Appl. Microbiol.* **2015**, *9*, 2725–2732.
48. Dineshkumar, R.; Narendran, R.; Jayasingam, P.; Sampath-kumar, P. Cultivation and chemical composition of microalgae *Chlorella vulgaris* and its antibacterial activity against human pathogens. *J. Aquac. Mar. Biol.* **2017**, *5*, 00119.
49. Hamouda, R.A.E.; Abou-El-Souod, G.W. Influence of various concentrations of phosphorus on the antibacterial, antioxidant and bioactive components of green microalgae *Scenedesmus obliquus*. *Int. J. Pharmacol.* **2018**, *14*, 99–107.
50. Ruffell, S.E.; Müller, K.M.; McConkey, B.J. Comparative assessment of microalgal fatty acids as topical antibiotics. *J. Appl. Phycol.* **2016**, *28*, 1695–1704. [CrossRef]
51. Schuelter, A.R.; Kroumov, A.D.; Hinterholz, C.L.; Fiorini, A.; Trigueros, D.E.G.; Vendruscolo, E.G.; Zaharieva, M.M.; Módenes, A.N. Isolation and identification of new microalgae strains with antibacterial activity on food-borne pathogens: Engineering approach to optimize synthesis of desired metabolites. *Biochem. Eng. J.* **2019**, *144*, 28–39. [CrossRef]
52. Santhakumaran, P.; Kookal, S.K.; Mathew, L.; Ray, J.G. Experimental evaluation of the culture parameters for optimum yield of lipids and other nutraceutically valuable compounds in *Chloroidium saccharophillum* (Kruger) comb. Nov. *Renew. Energy* **2020**, *147*, 1082–1097. [CrossRef]
53. Noaman, N.H.; Fattah, A.; Khaleafa, M.; Zaky, S.H. Factors affecting antimicrobial activity of *Synechococcus leopoliensis*. *Microbiol. Res.* **2004**, *159*, 395–402. [CrossRef] [PubMed]

MDPI
St. Alban-Anlage 66
4052 Basel
Switzerland
Tel. +41 61 683 77 34
Fax +41 61 302 89 18
www.mdpi.com

Applied Sciences Editorial Office
E-mail: applsci@mdpi.com
www.mdpi.com/journal/applsci

www.ingramcontent.com/pod-product-compliance
Lightning Source LLC
LaVergne TN
LVHW070614170726
843515LV00004B/74

* 9 7 8 3 0 3 6 5 3 7 9 7 9 *